7/1/93

Computer Methods in UV, Visible, and IR Spectroscopy

Computer Methods in UV, Visible, and IR Spectroscopy

Edited by
W. O. George
The Polytechnic of Wales, Pontypridd

H. A. Willis
Visiting Professor, University of East Anglia

ROYAL
SOCIETY OF
CHEMISTRY

British Library Cataloguing in Publication Data
Computer methods in uv, visible and ir spectroscopy.
1. Spectroscopy. Applications of computer systems
I. George. W. O. (William Owen) II. Willis. H. A.
535.840285

ISBN 0-85186-323-7

Published by The Royal Society of Chemistry,
Thomas Graham House, The Science Park, Cambridge CB4 4WF

Typeset by BookEns, Saffron Walden, Essex
and printed by Bookcraft (Bath) Ltd

Preface

The chapters of this book are essentially the texts of lectures given at the Royal Society of Chemistry Residential School at the Polytechnic of Wales in March 1989.

The lectures formed four subject sessions, namely:

(i)   Identification of materials from their IR spectra by computer band match and expert systems.
(ii) Data manipulation and 'combined techniques'.
(iii) Computer methods in quantitative analysis.
(iv) The status of user written software.

The talks were supplemented by practical sessions consisting of hands-on experiments and demonstrations.

The written texts in some instances follow closely the course notes, but in other cases have been extensively rewritten to cover questions raised and comments from participants.

The lecture on the data transfer system J CAMP-DX has been supplemented by some actual examples which formed part of the practical sessions while the course script on Chemometrics has been replaced with a script by a different author.

Opinions expressed in the scripts are those of the authors, and do not necessarily agree with the opinions of the editors who have examined the scripts for factual errors, tried to make them more easily readable, and have endeavoured to resolve any ambiguities.

H.A. Willis
W.O. George

Contents

Contributors

B.J. Clark, *Pharmaceutical Chemistry, School of Pharmacy, University of Bradford, Bradford BD7 1DP, UK.*

J.T. Clerc, *Universtat Bern, Pharmazeutisches Institut, CH-3012 Bern, Baltzerstrasse 5, Switzerland.*

J. Coates, *Nicolet Instrument Corporation, 5225-1 Verona Road, PO Box 4508, Madison, Wisconsin 53711-0508, USA.*

B. Davies, *Glaxo Manufacturing Services Ltd., Barnard Castle, County Durham DL12 8DT, UK.*

A.F. Fell, *Pharmaceutical Chemistry, School of Pharmacy, University of Bradford, Bradford BD7 1DP. UK.*

M.A. Ford, *Perkin Elmer Ltd., Post Office Lane, Beaconsfield, Buckinghamshire HP6 1QA, UK.*

A.S. Gilbert, *The Wellcome Research Laboratories, Langley Court, South Eden Park Road, Beckenham, Kent BR3 3BS, UK.*

H.J. Luinge, *University of Utrecht, Laboratory of Analytical Chemistry, Croesestraat 77a, 3522AD Utrecht, The Netherlands.*

P.S. McIntyre, *Department of Science and Chemical Engineering, The Polytechnic of Wales, Pontypridd, Mid-Glamorgan CF37 1DL, UK.*

S. Reade, *Department of Science and Chemical Engineering, The Polytechnic of Wales, Pontypridd, Mid-Glamorgan CF37 1DL, UK.*

H. Somberg, *Bruker Analytische Messtechnik GmbH, Wikingerstrasse 13, D-7500 Karlsruhe 21, W. Germany.*

I. Steer, *Philips Analytical, York Street, Cambridge CB1 2PX, UK.*

R.L. Tranter, *Analytical Development Division, Glaxo Operations UK Ltd., Barnard Castle, County Durham DL12 8DT, UK.*

J.H. van der Maas, *University of Utrecht, Laboratory of Analytical Chemistry, Croesestraat 77a, 3522 AD Utrecht, The Netherlands.*

P.S. Wilson, *Bio-Rad Microscience Ltd., Bio-Rad House, Maylands Avenue, Hemel Hempstead, Hertfordshire HP2 7TD, UK.*

CHAPTER 1

Introduction to Computer Methods

M. A. FORD

This paper is intended as a general introduction to the subject and is divided into two parts. Section 1 gives a broad historical background — not a detailed history — covering mainly information and events which influenced the author in his own work. The absolute accuracy of all dates is not guaranteed, but they are near enough to give a general appreciation of events (Table 1).

Section 2 covers the instrumental conversion of optical signals into digital spectroscopic data, since good appreciation of this basic step is considered to be helpful in relation to the subsequent computer processing of the data. It is illustrated with examples of ratio recording dispersive infrared, FTIR, and fluorescence spectrometers. The fact that the data can be influenced by a number of arbitrary factors is emphasized.

1 HISTORICAL INTRODUCTION

The earliest record known to the author of any digital data handling in spectroscopy, albeit manual, was by Michelson in 1892.[1] His raw digital data were based on a numerical visual assessment of the visibility of fringes in his interferometer or measured galvanometer deflections of a micro radiometer. However, even from this crude approach, he was able to perform manual* Fourier transforms and obtain recognizable spectra (Figure 1). The following are extracts from the 1892 paper.

'The principal object of the foregoing work is to illustrate the advantages which may be expected from a study of the variations of clearness of inter-

---

* The Fourier transformation was not a full 'forward' transform, but an iterative series of reverse transforms of estimated spectra to obtain the best fit.

[1] A. Michelson, *Philos. Mag.*, 1892, **34**, 280.

Table 1   *Some dates of interest*

1892	Michelson produces spectra from his interferometer.
1949	Felgett recognizes 'multiplex advantage'.
1953	Commercial FTIR proposed — proposal rejected.
1959	Bauman publishes 'a least squares method for multicompetent analyses'.
1959–	Many papers published on computer methods in spectroscopy.
1962?	First commecial FTIR (Far IR, step scan).
1964	Savitzky–Golay publish 'Smoothing and Differentiation of Data by Simplified Least Squares Procedures'.
1972	Intel introduces first microprocessor.
1972	First fast scan FTIR.
1975	First microprocessor controlled analytical instruments.
1975	First commercial computer spectroscopic data handling systems available.
1978	First fully microprocessor instruments, *i.e.* with all internal data handled digitally.
1978	First 'data stations'.
1979	First commercial spectral interpretation system.
1983	First chemometric software, *e.g.* 'Computerized Infrared Characterization of Materials' (CIRCOM).
1984–	More and better!

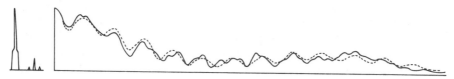

Figure 1   *Mercury green line spectrum and interferogram (observed and calculated)*[1]

ference-fringes with increase in difference of path. The fundamental principle by which the 'structure' of a line or group of lines is determined by this method is not essentially different from that of spectrum-analysis by the grating, both depending, in fact, on interference phenomena; but in consequence of the almost complete freedom from errors arising from defects in optical or mechanical parts, the method has extraordinary advantages for this special work.'

'The green mercury-line is one of the most complex yet examined. The constituent lines are nevertheless so fine that the interference-bands are frequently visible when the difference of path is over four tenths of a metre. The full curve in Figure 1 gives the results of observations corrected for personal equation, while the dotted curve represents the equation.'

From 1892 we jump to 1949 with the recognition by Felgett of the 'multiplex advantage' of interferometry — an improvement in signal to noise ratio proportional to the square root of the number of elemental signals measured simultaneously. The feasibility of a commercial infrared spectrometer utilizing this principle was considered as early as 1953, but abandoned because of the computational difficulties* at that time and it was not until about 1960 that the first commercial far infrared FT instruments were produced. These simply gave a digitized output of the interferogram on punched paper tape which had to be taken to a friendly owner of a main frame computer to be transformed. If you got your spectrum the next day you were doing very well!

Also, by 1960 Savitzky was working on computer processing of infrared spectral data.[2,3] His work was mainly related to peak detection and characterization to assist in both quantitative and qualitative analysis. The computer was a General Precision LGP30 using paper tape input and the operation was very tedious. Later, in 1964, Savitzky together with Marcel Golay published their paper on digital smoothing.[4] The importance of this can be appreciated by the fact that, over the intervening years, it has become one of the most frequently cited papers ever published.

The author's own earliest involvement in digital data handling was on computer simulations of the characteristics of the analogue filters which were then being used in instruments. The work, in the late sixties and early seventies, was done on a Honeywell 3/16 computer and, although tediously slow, gave some interesting results. At that time it was believed that it would never be possible to use digital filtering in real time as the computational requirements would be too great.

In 1972 the author was introduced to the concept of a broad, multitechnique computer spectral handling system — more than a straight forward library search system — by Professors Simon and Clerc of the ETH, Zurich. Working with many others, this led eventually to both a library search and a spectral interpretation system in 1979. Although a number of library search systems,[5-8] including manual card index type, had been introduced earlier, it is believed that this was the first commercial interpretation system. In the same year (1972) there was another very significant event, Intel produced the first microprocessor! However, it was another 3 years before the first microprocessor controlled instrument was introduced and this used a microprocessor only for control, not for data handling.

It was at least another 2 years, *i.e.* 1977, before a microprocessor was first used in an instrument to handle the data in real time. However, routine digitization of instrument data had started much earlier, led, in fact, by gas chromatography, where it became standard practice to perform digital integ-

---

* The fast Fourier transform algorithm had not been devised at that time and the full transform had to be performed — a heavy computational task even today.

[2] A. Savitzky, *Anal. Chem.*, 1961, **33**, 25A.
[3] A. Savitzky, *Anal. Chem.*, 1989, **61**, 921A.
[4] A. Savitzky and M.J.E. Golay, *Anal. Chem.*, 1964, **36**, 1627.
[5] D.H. Anderson and G.L. Covert, *Anal. Chem.*, 1967, **39**, 1288.
[6] G.A. Massios, *Am. Lab.*, 1971, **3**, 55.
[7] D.S. Erley, *Appl. Spectrosc.*, 1971, **25**, 200.
[8] R.W. Sebesta and G.G. Johnson Jnr., *Anal. Chem.*, 1972, **44**, 260.

ration before 1957. Nearly all the early spectroscopy work was done using external A to D conversion of analogue signals from the spectrometers,[9] often recording the data on paper tape for subsequent use. In addition to the library search work mentioned earlier, a number of centres were working on such things as spectral difference, spectral formatting procedures, spectral deconvolution,[10–16] and various quantitative methods. In connection with the latter it is worth noting that the least squares method was published by Bauman in 1959,[17–19] but it had been used on mass spectral data at ITT as early as 1956.

Perkin-Elmer introduced its first computer spectral data handling package in 1975; it included such features as spectral difference, averaging, formatting, and transmittance to absorbance conversion. It is interesting to note that a least squares curve fitting quantitative analysis programme was also available for the same mini computer at that time, but it was to be another 10 years before this technique started gaining general acceptance.

The next major step was the introduction, in 1978, of the first 'Data Stations', in a sense the forerunners of the modern PCs, together with macro languages to enable the user to control the instrument and sequence a number of operations on the data with a single command, a very fundamental step towards automation. However, it was probably the advent of the first fast scanning FTIRs which gave the greatest stimulus to computer handling of spectroscopic data. The reason for this was simply that the date was inherently digital and the computer was already there!

The next significant development, in the author's view, was of the more statistical type of approach to extracting information from spectroscopic data — the field which has now become known as chemometrics. A very early practical user of this type of approach was Peter Fredericks of Broken Hill Proprietary in Australia who, with co-workers, had devised their 'CIRCOM' [20–22] approach for obtaining measurements of physical and chemical properties of materials from their IR spectra. It was applied initially to the determination of the properties of coal — such things as specific energy and total ash content — and later extended to other minerals. The technique is widely used throughout BHP. This type of approach has enormous potential and effectively adds another dimension to the information which can be extracted from our spec-

[9] R.N. Jones, *Pure Appl. Chem.*, 1969, **18**, 303.
[10] A.L. Khidir and J.C. Decius, *Spectrochim. Acta*, 1962, **18**, 1629.
[11] R.N. Jones, K.S. Seshadri, N.B.W. Jonathan, and J.W. Hopkins, *Can. J. Chem.*, 1963, **41**, 750.
[12] R.N. Jones, R. Venkataraghavan, and J.W. Hopkins, *Spectrochim. Acta, Part A*, 1967, **23**, 925.
[13] R.N. Jones, R. Venkataraghavan, and J.W. Hopkins, *Spectrochim. Acta, Part A*, 1967, **23**, 941.
[14] G. Horlick, *Appl. Spectrosc.*, 1972, **26**, 395.
[15] G. Brouwer and J.A.J. Jansen, *Anal. Chem.*, 1973, **45**, 2239.
[16] F. Zenitani and S. Minami, *Jpn. J. Appl. Phys.*, 1973, **12**, 379.
[17] R.P. Bauman, *Appl. Spectrosc.*, 1959, **13**, 156.
[18] J.C. Sternberg, H.S. Stillo, and R.H. Schwendeman, *Anal. Chem.*, 1960, **32**, 84.
[19] H.A. Barnett and A. Bartoli, *Anal. Chem.*, 1960, **32**, 1153.
[20] P.M. Fredericks, P.R. Osborn, and D.A.J. Swinkels, *Fuel*, 1984, **63**, 139.
[21] P.M. Fredericks, J.B. Lee, P.R. Osborn, and D.A.J. Swinkels, *Appl. Spectrosc.*, 1985, **39**, 303.
[22] P.M. Fredericks, P.R. Osborn, D.A.J. Swinkels, *Anal. Chem.*, 1985, **57**, 1947.

troscopic data by computers! Again, acceptance of the approach has been slow, but it is certainly now gaining momentum.

To conclude the first section of this paper it must be stressed that computer methods can never add to our 'raw' data (they can, and often do, detract from it). They can only convert it into a more easily understandable form, whether that be a smoothed spectrum, a chemical identification or a quantitative analysis. Figures 2 and 3 are a light hearted illustration of 'conversion into a more easily understandable form'. The series of dots of Figure 2 become the easily recognized message of Figure 3 when joined by computer using an appropriate interpolation algorithm.

2 SIGNAL DIGITIZATION

This section of the paper is concerned with the generation of the original digital signals from our spectrophotometers — the so called raw data. The words 'so called' are used to emphasize the fact that the data are never absolute in the sense of being free from some arbitrary constraints. This is explained in more detail in relation to the specific examples given later.

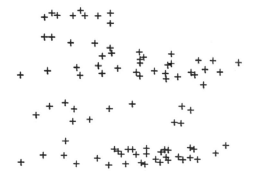

Figure 2

Figure 3

Table 2   *Rules for Analogue to Digital conversion*

A   The smallest measurable increment must be less than the noise.
B   The sampling rate must give at least two samples per cycle of
        the highest *noise* frequency.
C   The dynamic range must be sufficient to accept the largest
        signal.

Obviously, the critical step in the generation of raw data is the conversion of
the analogue output from the detector into digital form. Table 2 lists 3 simple
rules which must be observed in order to carry out this step without loss of any
information. Rules A and C are reasonably obvious but Rule B requires some
explanation. Inherently there will always be noise present in any signal   at
higher frequencies than the observed signal. If such noise is to be properly
averaged then the requirment of Rule B must be observed.

Example 1   Ratio Recording Dispersive IR

Figure 4 illustrates the data path through a ratio recording dispersive infrared
instrument. The detector measures signals representing sample, reference,
sample blank, and reference blank in a mechanically defined sequence. Even
at this point, there is an arbitrariness in the data concerned with the nature of
the transition from one signal to the next. The detector, of course, has a
response time constant which adds a further arbitrary factor.
    The signal is amplified and integrated for the duration of each of the optical
segments before A to D conversion, the integration being a good way of com-
plying with Rule B. For each cycle of the optical chopping system, the 4 signals
are stored in a running buffer and data from 3 complete cycles are used to calculate 'corrected' sample and reference values. This correction is required to
deal with two effects: first, the slow response of the detector means that the
integral determined for each optical segment contains information from
earlier segments and, second, if the instrument is scanning continuously the
measurement of the various segments will not occur at the same wavenumber.
The actual form of the equation, which is quite complex, need not concern us,
except to note that it contains arbitrary factors to allow for individual detector
response characteristics.
    After the calculation of these 'corrected' sample and reference signals, the
signal is frequently block averaged before ratioing. This is necessary when
noise forms a significant percentage of the reference signal — under such circumstances the average of ratios is not the same as the ratio of averages, the
latter being correct. The result of the ratio can be considered as 'raw' spectral
data but, depending on scanning conditions, this data may contain noise at a
much higher frequency than the most rapidly changing possible spectral data.
Savitzky–Golay filtering is used to minimize this higher frequency noise with
the filter width matched to the scanning condition. The filtered spectral data is
what is normally output for use in the computer methods which are the sub-

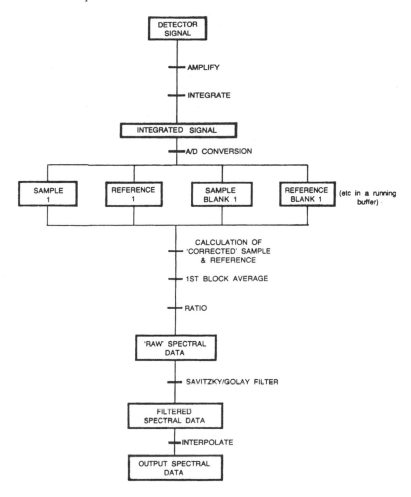

Figure 4   *Dispersive IR (ratio recording) data path*

ject of this book. The final process of interpolation is used if we require data at a different spectral interval.

If 'raw data' is required a valid case can be made for taking the data before the Savitzky–Golay filter, particularly if measurements are to be made of the noise level itself, but using data any earlier in the sequence would be disastrous because of the large number of arbitrary factors involved in it.

Example 2   FTIR

Figure 5 shows the corresponding data path for an FTIR instrument. The fact that the sample and reference (background) signals are generated in separate

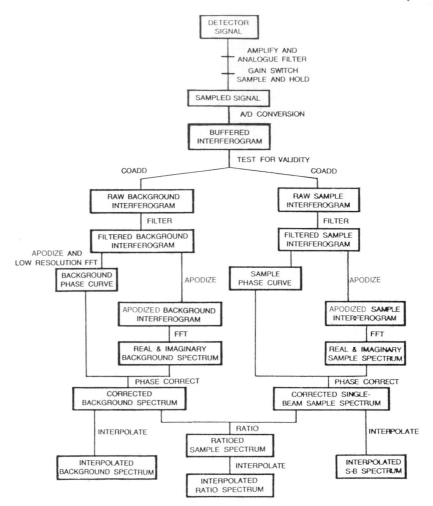

Figure 5

scans gives us some degree of simplification. In the FTIR case the data sampling rate is such that it is not practical to use integration to comply with Rule B and, therefore, it is necessary to use a suitable analogue filter to, as far as possible, eliminate the higher frequencies from the signal without restricting the signal itself. The filtered signal, which may be gain switched to extend the range of the A to D converter, is sampled at specific points and converted to digital form. The gain switching can be justified against rules A and C since, in most real cases, there is some degree of variation of noise with signal, so that when the signal is very large the smallest measurable increment can also be larger.

   The complete digital interferogram is held in a temporary buffer and only co-added to other scans if sensors in the instrument have not detected an abnormal situation, such as a strong electrical or mechanical pulse.

Before performing the Fourier transform of the interferogram it is necessary to use a digital filter to eliminate signal frequencies which would result in aliasing, *i.e.* high wavenumber signals appearing as if they were at a lower wavenumber. It should be noted that the analogue and digital filters work in combination and the digital filter could be avoided by the use of suitable analogue filtering. However, much better, overall, characteristics can be obtained by the use of both.

A further step before Fourier transform is apodization, which refers to the attenuation of the signals at longer path differences by some arbitrary function to minimize the artifacts created by the non-infinite nature of the transform. For brevity, this will not be explained, but it should be remembered it is another arbitrary choice, along with the filter characteristics.

Yet another arbitrary choice is the nature of phase correction used. If an interferometer were perfectly symmetrical then it would produce a symmetrical interferogram and no phase correction would be necessary. In practice there is always some asymmetry resulting in both real and imaginary (cosine and sine) terms in the spectrum. In principle the phase correction performs a mathematical rotation of the data to give the largest positive value — an alternative is to square the real and imaginary spectral data, add and take the square root.

After phase correction it is necessary to ratio the sample and background spectra, that is divide the 'sample' energy by the background energy at each wavenumber. This removes the instrument response function and, as in the dispersive case, the data may be interpolated to a more convenient spectral interval.

When the spectrum is generated using the fast Fourier transform algorithm the fundamental spectral interval is defined by the path difference interval at which the original interferogram is sampled. This, in turn, is normally directly related to the frequency of a helium–neon laser used as an internal reference, resulting in a series of wavenumber values which are neither integer numbers nor separated by integer numbers.

Because of this, the wavenumber values of data points are normally quoted to 3 places of decimals which can lead to a very misleading impression of accuracy.* This is particularly relevant when a peak position is given as the wavenumber value of the nearest data point: *e.g.* a value of 1621.469 cm^{-1} could often refer to a peak anywhere between 1621 cm^{-1} and 1622 cm^{-1}. In these circumstances there are considerable advantages in interpolating the data to exact wavenumber values to avoid both giving misleading information and potential problems in data transfer systems. The interpolation referred to is not a simple linear interpolation, but a highly accurate 'sinc interpolation' over a large number of data points.

---

* The often quoted wavenumber accuracy of FTIR instruments of ±0.01cm^{-1} is only achieved under very restricted and carefully controlled conditions. In normal use an uncertainty of several tenths of a reciprocal centimeter is more typical.

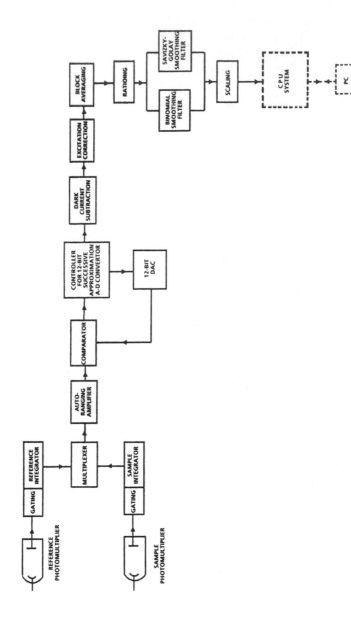

Figure 6 *Data system of a pulse source fluorescence spectrometer*

Example 3 Fluorescence Spectrometer

The data system of a pulse source fluorescence spectrometer is shown in Figure 6. In a similar manner to the dispersive infrared instrument, signal integration is used to comply with Rule B but, in this case, separate detectors are used for sample and reference. In addition gating is used to eliminate noise during the relatively long interval between source pulses. Although the total dynamic range from the noise level to the largest signal approaches 18 bits, *i.e.* 250 000:1, it is only necessary to use a 12 bit A to D converter together with gain ranging, since the signal to noise ratio never exceeds 12 bits or 4000:1. The reason for this is that the noise is almost entirely shot noise, *i.e.* due to photon statistics, and therefore varies as the square root of the signal.

Both detectors are, in practice, gated not only during each source pulse, but also for a brief period between pulses in order to measure dark current which is digitally subtracted. Again, block averaging is used before ratioing and a choice of Savitzky–Golay or binomial filters provided.

An additional step in the data sequence compared to the infrared instruments is emission correction, which refers to multiplying the signal by a previously measured wavelength dependent function to give a constant quantum response.

For this instrument there are fewer arbitrary factors, but, even so, there are still the detector and preamplifier response, the duration and timing of the gating, the form of the excitation correction, and the nature of the filtering.

The description of the above examples has necessarily been fairly detailed in order to give some appreciation of the operations which will have already been performed on the data *before* it is used in the computer methods which are the subject of this book. It is hoped that an understanding of the need for these operations and their somewhat arbitrary nature will be of some help in assessing and interpreting data.

Acknowledgement. The author wishes to acknowledge the late Larkin B. Scott for the provision of Figures 2 and 3.

CHAPTER 2

Techniques for Searching Data for Qualitative Analysis

J. T. CLERC

1 INTRODUCTION

This chapter is intended to give the reader a general introduction to the spectra interpretation problem. In the first part a generalized overview of the theoretical background is presented in the context of the practical analytical laboratory with emphasis placed on the library search approach. The more sophisticated so-called artificial intelligence techniques are covered in another chapter. The goal of this chapter is to make the reader acquainted with the general theory of library search systems in order to understand what a library search system can do and what it cannot do. Furthermore, the problem of evaluating different library search systems, *i.e.* of measuring the relative performance is addressed. Finally, areas where insufficient knowledge blocks further progress and thus need more basic research are indicated.

2 THE SPECTRA INTERPRETATION PROBLEM

2.1 Theoretical overview

Analytical spectroscopy for qualitative analysis is based on the fact that there is a relation between the chemical structure of a compound and its molecular spectra. In abstract mathematical form this can be written as the spectrum being a function of the structure:

$$\text{'spectrum'} = F\,(\text{'structure'})$$

If this function F were exactly known and if we were able to invert this function to F^{-1}, then we could apply F^{-1} to the spectra of the compounds to identify and directly get their structures.

$$\text{`structure'} = F^{-1} \text{(`spectrum')}$$

However, the function F is neither known with sufficient precision nor is it invertible. Thus, a simplified approach is generally used. The function F is partitioned into partial functions f_i which operate on partial structures and partial spectra:

$$\text{`partial spectrum } i\text{'} = f_i \text{(`partial structure } i\text{')}$$

These partial functions f_i are selected to be invertible. Thus, we can apply their inverses to partial spectra of the sample at hand to get its partial structures:

$$\text{`partial structure } i\text{'} = f_i \text{(`partial spectrum } i\text{')}$$

However, by splitting up the global function F into partial functions f_i we lose the context information, the inverted functions f_i^{-1} become ambiguous. Thus, whereas a given partial structure gives one unique set of spectral properties, the application of the inverted function f_i^{-1} becomes multivalued. Non-spectroscopic information is required to select the correct solution among the possible ones.

2.2 Practical View

The abstract theory given above is easily translated into the practical world of the analytical spectroscopist. The spectrometer may be looked upon as an analog computer which evaluates the function F, converting structural data into spectral data. The partial functions f_i are represented in correlation charts and tables, where the spectral data expected for various structural elements are tabulated. These tables and charts can be accessed from both sides, they represent invertible functions. For one given structural element, one set of spectral data is given. For a given set of spectral data however, many different structural elements are generally possible. Thus, the primary result is a set of partial structures containing many noise fragments (*i.e.* fragments not present in the target structure). From this collection of fragments the analyst tries to assemble full structures. This process is controlled by non-spectroscopic knowledge, in particular by the rules of structural chemistry and by knowledge about the sample's origin. Assembling fragments into a full structure recovers the context information lost in the previous step. Thus, prediction of the spectral data expected for a given trial structure is inherently more precise than predicting structural data from spectra. Therefore, the analyst checks his hypothetical structures by predicting their spectral data. Besides correlation tables and charts he may also use other non-invertible representations of partial functions f_i such as additivity rules, or even representations of the full function F, *i.e.* reference spectra. Only those structure proposals for which the predicted spectral data is closely similar to the sample spectrum will be retained.

2.3 General Algorithm for Spectra Interpretation

The procedure described above may be formalized into the following general algorithm, which is closely followed by all systems for computer supported spectra interpretation:

(i) *Spectra Recording*. Record the spectral data for the sample at hand.

(ii) *Partial Structures*. For the spectral data obtained in step (i) prepare a list of possible partial structures by using suitable representations of the functions f_i^{-1} relating partial structures to partial spectra, *i.e.* correlation charts and tables, and rules.

(iii) *Consistent Sets*. From the list assembled in step (ii) select all possible sets of structural elements possibly present simultaneously in the sample at hand.

(iv) *Structure Assembly*. From all sets generated in step (iii) produce all possible full structures (tentative structures) which are compatible with the rules of chemistry and not at variance with the non-spectroscopic evidence available.

(v) *Spectra Prediction*. Predict the spectral data for the next tentative structure generated in step (iv), using correlation charts and tables, additivity rules, and reference spectra. If there are no more tentative structures then stop. The list assembled in step (vi) now contains all structures possible for the sample at hand.

(vi) *Spectra Comparison*. Compare the data predicted in step (v) with the data recorded in step (i). If they are closely similar, keep the respective tentative structures as a possible solution in a list, else drop it. Then continue with step (v).

Systems using the artificial intelligence approach (the term Expert System currently becoming more popular) implement one or more of these steps explicitly in a computer program. The various systems so far described in the scientific literature vary greatly in the degree of sophistication applied to the implementation of the different steps. However, the basic strategy remains the same for all of them. Chapter 3 will specifically deal with such systems. Library search systems also follow the same pattern, however in a very much simplified way.

3 THE LIBRARY SEARCH APPROACH

3.1 Overview

When we do a library search, we proceed in the following way. First we record the spetra of the sample at hand. The spectra of the sample are then compared to all reference spectra in the reference library. If there is a sufficiently close match between the unknown sample's spectra with one of the reference spectra, the chemical structure of the respective reference compound is believed to be similar to the sample's chemical structure. This seemingly simple process follows closely the steps given in the general algorithm. To gain insight into the potentials and limitations of the library search approach a somewhat more detailed system analysis is helpful.

In a library search system the function-F is not partitioned explicitly into partial functions. The function F is rather used as a pointwise representation by a collection of reference spectra, very much like a table of logarithms pointwise represents the function $y=\log (x)$.

With regard to the general algorithm given above, steps (ii) and (iii) (partial structures and consistent sets) are not performed explicitly. For step (iv), all structures represented in the library are selected. As the library contains the data for real compounds, the rules of chemistry are automatically followed. Step (v) spectra prediction becomes trivial, as the full data for all tentative structures are available. The only critical step is step (vi), the spectra comparison.

3.2 The Basic Model

The most important thing to realize when working with library search systems is that when we compare spectra we really mean chemical structures. We tacitly assume that similar spectra imply similar chemical structures. If in a library search system this hypothesis is not adequately met, the respective system will be useless for practical applications. This is a necessary but not sufficient condition for a library search system to be useful. Thus, in order to understand library search, one has to understand the comparison process. Thus the central hypothesis of library search is:

if the spectra are similar then the chemical structures are similar.

Whether this statement is true or not obviously depends on what is meant by 'similar'. We consider two objects to be similar if they are identical in many aspects believed to be relevant in the given context. Similarity thus requires that the objects to be compared are characterized by many different features which can be either identical or different. A simple measure of the similarity between two objects is the number of corresponding features in which they are identical. Thus, a chemical compound and its spectra are characterized by a set of structural descriptors and by a set of spectral descriptors. One may now set up two independent multidimensional spaces for structures and spectra respectively, where one descriptor is assigned to each coordinate axis from the respective set. Every conceivable chemical compound can be mapped into a point in each of these multidimensional spaces, using the values of the respective descriptors as coordinate values. The two spaces are referred to as the structure space and the spectra space respectively.

Comparing the spectra of two compounds amounts to measuring the distance between their respective points in the spectra space. The more close they are, the more similar are the two spectra. If the two points coincide, the spectra are considered as identical. The library search hypothesis now requires that points close together in the spectra space are also close together in the structure space, *i.e.* the two spaces should have the same (mathematical) structure.

The spectral features defining the spectra space and the distance measure to be used in this space are an integral part of any library search system. The goal in selecting the descriptions and in defining the distance measure is to model in the spectra space the (mathematical) structure of the structure space as closely as possible. Similarity in the structure space, however, is not defined and depends on the problem to be solved and on the user and his or her preferences. Thus, no library search system can work well for all applications.

3.3 Feature Selection

The spectral feature selected to describe the spectra and to span the spectra space determine how well the spectra space will conform to the structure space. One wants the spectral features to be sensitive to differences in the chemical structure, but insensitive to technical and instrumental conditions.

In some applications one may assume that the unknown samples are represented in the reference library. This is often the case in environmental analysis and in government laboratories, where the question is whether the sample is a compound specified in a (limited) list. Here, the degree of structural sensitivity of the features selected is of no great concern. The system is expected to identify all compounds in the list. If a sample is another compound, the answer 'not identical to a reference compound' is interpreted as 'not on the list' and is accepted as such. Applications of this type are referred to as identity search.

However, if the question is 'What is it?', then feature selection becomes much more difficult. If the sample is not identical to one of the reference compounds, the user expects to be presented with a set of model compounds similar to the unknown and uses the similarity measure (distance in the spectra space) as an estimate of structural similarity (distance in the structure space). To get reasonable correlation between the two similarity measures the spectral features have to be selected with care. This type of application is referred to as similarity search.

3.4 Size and Contents of the Library

All library search systems have one fundamental limitation. The set of answers they can provide the user with is delimited by the contents of the reference library. If, for a given unknown, no suitable reference compound is part of the library, no useful answer can ever result, independent of all other factors.

Thus, size and contents of the reference library are of fundamental importance for the performance of the system.

For an identity search system it is obvious that the library has to include all compounds to be identified. The larger the library, the more compounds we can successfully deal with. However, the other aspects of the performance are not increased by additional compounds. On the contrary, search time and thus costs all increase with the size of the library.

For similarity search systems, suitable reference compounds for all conceivable unknowns are required. This seems to call for a comprehensive library containing the spectra of all known chemical compounds or at least as many as are available. However, a moments reflection shows that this is not the optimal solution. For each compound type one needs just one or maybe a few references. Large sets of closely related reference compounds will, in a true similarity search system, not increase the performance. On the contrary, retrieving an excessive number of closely similar references for a given sample will just increase the output volume without providing additional information. It is only the first reference compound in a group of related compounds which matters. The other references, being very similar to the first one, just repeat the same message over and over again.

What one really needs can be clearly stated within the space model: the spectra space has to be filled sparsely but evenly with reference compounds, to provide some close neighbours in every section of the space. The correspondence between the spectra space and the structure space insures that in this case the structure space will also be adequately populated. The population density should reflect the relevance of the respective space sector to the user. In his fields of interest the average number of references may be higher than elsewhere, in order to give a higher resolution. Thus, an optimal library will consist of two parts. A general part, which provides a few prototype compounds for every compound class, and one or more special parts, documenting the user's field(s) of interest with enhanced resolution.

To ensure wide coverage of all fields and branches of chemistry, the general part of a reference library will probably have to be bought from an outside source. The special parts, however, have to be assembled by the user himself. All the spectra recorded in his laboratory are important in this context, even if the respective compounds have no other relevance. The fact that the respective sample has been submitted for analysis is proof that it is relevant to the analyst. Thus, all spectra recorded in the user's laboratory are candidates for inclusion in the reference library. However, to limit the size of the reference library only the spectra of compounds from classes not yet adequately represented in the library should be added.

4 EVALUATION OF THE PERFORMANCE

The performance of a library search system includes many different aspects. First of all, we expect a powerful library search system to be able to retrieve from the reference library compounds identical to the sample, if such compounds exist in the library. If no compound identical to a given sample exists in the library, the system should provide the user with reference compounds structurally similar to the unknown. It should be able to do so regardless of the technical parameters used when recording the respective spectra. To each reference retrieved a reasonable measure of similarity to the unknown should be put out. These similarity values have to inform the user whether a given reference compound can be assumed to be structurally identical to the sample with high probability or whether it is only similar. In the latter case the similarity measure should give a reasonable estimation of the structural similarity. Furthermore, we expect the library search system to be fast, to be easy to use, and to present the results in an easily interpreted form. Finally, there have to be program modules which allow for easy maintenance of the system library. In particular it should be easy to add, delete, and edit entries. Evaluating a library search system amounts to assigning values (measured at least on an ordinal scale) to all these aspects of performance. In the present chapter no attempt is made to discuss the aspects which depend to a large extent on the user's personal taste, *i.e.* the presentation of the results and the quality of the user interface. The only topic to be discussed is how to estimate the quality of the results from a chemical point of view.

4.1 General Considerations

The most important qualifier for a library search system is how well it maps spectral similarity into structural similarity. The fundamental problem here is that presently no generally accepted similarity concepts for chemical structures exist. The chemist's notion of structural similarity is strongly coined by traditional views (functional groups, skeletons, compound classes) and by the problem he currently studies. Furthermore, different spectroscopic techniques focus on different structural entities and are thus inherently biased towards certain aspects of structural similarity. Thus, no generally applicable procedure can be given.

A second problem arises from the fact that the composition of the reference library is of paramount importance. The set of possible answers is limited to the chemical structures represented in the library. A system to be tested can produce a useful answer only if its library contains at least one suitable reference compound for the test sample. Thus, an unlucky choice of test samples can shift an otherwise excellent system to the bottom of the list.

4.2 Test Procedures

Both these problems can be solved by using a standardized test library to be used by all systems to be compared. The user should carefully select a set of about ten to twenty test compounds and record their spectra under strictly routine conditions. It is important that the user selects these compounds to be truly representative for the compound classes he predominantly deals with in his practical work. This ensures that the test is performed in the regions of the spectra and structure space most important to the user. The test compounds are grouped in subgroups (preferably triplets) containing at least one pair of (structurally) highly similar compounds and one pair with moderate structural similarity, the degrees of structural similarity being subjectively judged by the user in the context of his daily routine work. With this choice the user implicitly defines his concepts of structural similarity. Between the subgroups, however, the structural similarity should be low. One or a few compounds should be duplicated, *i.e.* their spectra should be recorded twice under different conditions (different matrix and/or different instrument settings).

The test spectra are then assembled into a mini reference library to be used by all systems to be tested. The same spectra will also be used as test samples. Each sample compound is submitted to the system as an unknown, and for all references the similarity computed by the system is noted. The data are best organized into a similarity matrix. Each unknown is assigned to a row, the columns are assigned to the references, and the members of subgroups of mutually similar compounds are kept together.

This procedure makes sure that for each test sample there is at least one structurally identical compound in the library. Furthermore, for each sample there are one or more references which the user considers to be structurally similar. The duplicates will give information on how well the systems under test can handle variations due to technical artifacts.

4.3   Data Interpretation

From the similarity matrix a wealth of information regarding the behaviour of the system under test may be obtained. The diagonal elements correspond to comparing two identical spectra. Their values will correspond to the highest possible similarity, namely to perfect identity. If this is not the case, the mathematics of the system are wrong, the system behaves irrationally and further tests are unnecessary.

The rows and the columns for the duplicated samples show how well the system can deal with instrumental and technical variations. In the ideal case, the respective rows and columns should be identical. (However, a row is not necessarily identical to its corresponding column, the similarity matrix not necessarily being symmetrical). Any differences between corresponding values are due to the fact that the system misinterprets technical variations as being caused by structural differences. In a real system this is unavoidable. The root mean square difference between corresponding values is a measure for the amount of noise generated by different registration parameters and/or different matrices. Differences in similarity are meaningful only if they are significantly greater than this value.

Off-diagonal elements for duplicates should exhibit similarity values not significantly smaller than the diagonal elements. Failure to do so indicates that the system, being too sensitive for instrument settings and matrix effects, is unable to reliably recognize the identity of compounds.

In the submatrices corresponding to subgroups of mutually similar compounds all off-diagonal elements should be significantly smaller than the values for duplicate compounds to indicate to the user that sample and reference are not identical. If this condition is not met, then the system has difficulty in discriminating between true identity and high similarity. The values should, however, be significantly greater than the values of off-diagonal elements not being in a submatrix for a subgroup, to indicate that the respective compounds are definitely more similar than two compounds picked at random. There are two ways to violate this requirement. There can be some elements in the remainder of the matrix which are exceptionally high. As long as there are only a few such entries, this is of not great concern. It is just an indication that the system's inherent similarity concept for the structures includes some aspects neither obvious nor relevant to the user. The other case, some elements in the submatrix being too small however, has to be taken seriously. It indicates that the system's similarity measure does not consider certain aspects definitely important to the user. Finally, the sequence of similarity values should reflect the different degrees of structural similarity expected by the user. However, only differences clearly above the noise level are meaningful for ranking.

The information extracted from the similarity matrix as stated above will give a reasonably clear indication as to whether the system under test uses basically the same similarity concepts as the user and whether its sensitivity to structural variations and its insensitivity to instrumental and technical parameters is low enough to give sufficient discrimination between the cases of identity, of high similarity, and of no similarity.

One important point has to be kept in mind, however. The above qualifiers are to a large extent subjective. They measure the degree of correspondence between the user's and the system's concepts of structural similarity. The system's similarity concepts are defined implicitly by the designer, mainly by the spectral features selected for comparing spectra. They define the overall mathematical structure of the spectra space and its mapping into the structure space. By selecting groups of (in his or her opinion) structurally similar test compounds the user defines (again implicitly) the local mathematical structure of the spectra space. Thus, a given system not exhibiting top performance only proves that the system under test does not use structural similarity concepts similar to the user's. This may be due to a poor design of the system or due to the user's expectations being too specialized or irrational. Failure to perform as expected thus disqualifies a system only within the context specified by the test compounds. For other applications the respective system may well be the perfect choice.

Analysis of the similarity matrix can also supply information as to the search strategy implemented in the system. Under the assumption that the test compounds have been well selected, the following procedure reveals the respective information. First, all entries relating to identical spectra (diagonal elements only, not duplicates) are removed from the matrix. The remaining entries are ranked in decreasing order of similarity. Then one prepares a plot of similarity *versus* rank. This will give a curve which starts at rank 1 with a very high value for similarity and which subsequently drops to the lower similarity values for the later ranking pairs. The general shape of this curve is related to the search strategy employed by the system.

A similarity search system (as opposed to an identity search system) is expected to be able to retrieve reference compounds structurally similar to an unknown sample. Furthermore, the similarity measure should be able to discriminate between different degrees of similarity. Thus, the data set containing user selected pairs of high but variable similarity, the first part of the curve should be a smoothly declining curve, preferably almost linear. The length of the section to be considered is given by the number of off-diagonal elements in submatrices corresponding to the groups of structurally similar compounds. The shape of the remaining curve is of no concern. (Nobody cares how useless a useless reference is.) In an identity search system one expects positive identification of identical pairs, but places no particular emphasis on the similarity of non-identical pairs. The rank *versus* similarity curve thus starts with a short almost horizontal section, whose length is given by the number of duplicate pairs in the test set. The curve should then drop sharply and level off slowly. Of course, there is a gradual transition between the two pure strategies, resulting in curve shapes somewhere between these extremes. However, a rough estimate of the search strategy is generally possible.

An important part of the search strategy cannot be evaluated from the data in the similarity matrix. All similarity measures employed in library search systems are in some way based on the number of elementary spectral attribute states (the same value) in the two spectra to be compared. This number is set in relation to the total number of attributes considered. If all attributes do have the same state, the system considers the two spectra to be identical. Thus, the degree of similarity is measured as the number of attribute states identical between the two spectra compared, relative to the number of attribute states identical between one of the two spectra compared to itself. Whether the spectrum of the unknown or the spectrum of the reference is used as the base can lead to large differences in the system's behaviour. This is most easily explained using a strongly simplified mathematical model.

Let the set of features present in the unknown sample X be G_X and the corresponding set in the reference G_R and let the intersection of G_X with G_R be the set G_{XR}, the set of the features the two compounds X and R have in common. The number of members in the three sets G_X, G_R, and G_{XR} shall be designated as T_X, T_R, and T_{XR} respectively. A very crude measure for the similarity between X and R is then given by T_{XR}. To normalize this result it is divided either by T_X or alternatively by T_R. In the first case, where the unknown sets the standard, the similarity becomes independent from the number of features present in the reference only. Thus, the system does not penalize excess features in the reference. It just tries to make sure that the highest possible number of features present in X are also present in R. There is thus a tendency to retrieve references which are 'too big'. Each reference tries to represent all features of the unknown as completely as possible, the unknown tends to be a subset of the reference. This approach is referred to as a 'forward search'. It is particularly appropriate for similarity search of unknowns which can be assumed to be reasonably pure.

If the reference sets the standard, excess features in the unknown go unpenalized. The system prefers references which are 'too small'. Each reference tries to be completely represented in the unknown, it tends to be a subset of the unknown. This strategy is referred to as 'reverse search'. It has its main application with samples suspected or known to be mixtures, because it gives the user a chance to identify components of the mixture.

Of course, forward and reverse searches produce different results only if the spectra of the unknown and of the references are treated differently. If this is the case, a non-symmetrical similarity matrix results. The mathematical model given above is very much simplified. A more detailed analysis shows that the resulting search strategy further depends on whether the case of a given feature being present in both spectra is valued differently from the case where the same feature is absent in both. If this is the case a non-symmetrical similarity matrix results even if the two spectra are not treated differently. At present, however, there is no method known to determine from the similarity matrix which search strategy is used by a given system. The only thing one can say is that systems producing a symmetrical similarity matrix either treat unknown and reference alike or use a balanced search strategy exactly halfway between the two extremes.

5 OUTLOOK

The philosophy behind any library search system for the interpretation of molecular spectra is based on the hypothesis, that similar spectra imply similar chemical structures. A library search system is useful in the real world of the analytical chemist only if it makes this hypothesis come true. In order to build better library search systems, one needs similarity measures for both spectra and structures. The designer of a library search system decides upon the similarity for spectra. The similarity measure for structures, however, is defined by the user and depends on the problems he has to solve. In some cases he may be predominently interested in references having the same functional groups as his unknowns and does not care a lot about the skeleton. In other applications, however, he may place main emphasis on the skeleton. Consequently there will never be one single and universally applicable similarity measure for chemical structures. Presently there is none at all. The designer of a library search system is therefore faced with the impossible job of optimally mapping a similarity measure for spectra onto an unknown and undefined similarity measure for structures. Currently he or she has no other choice than to rely on guesswork. The development of even a poorly performing similarity measure for chemical structures would allow for the application of formalized mathematical optimization methods and would almost immediately lead to better library search systems. Thus, future research in this field should concentrate on chemical structures rather than on spectra.

6   BIBLIOGRAPHY

J.T. Clerc, E. Pretsch, and M. Zürcher, *Mikrochim. Acta*, 1986, **II**, 217.

J.T. Clerc, in 'Computer-Enhanced Analytical Spectroscopy', ed. Henk L.C. Meuzelaar and Thomas L. Isenhour, Plenum Press, New York, 1987, p. 145.

J.T. Clerc, in 'Research Instrumentation for the 21st Century', ed. Gary R. Beecher, Martinus Nijhoff Publishers, Dordrecht, 1988, p. 403.

M. Zürcher, J.T. Clerc, M. Farkas, and E. Pretsch, *Anal. Chim. Acta*, 1988, **206**, 161.

CHAPTER 3

Expert Systems for Automated Interpretation of Molecular Spectra

H.J. LUINGE AND J.H. VAN DER MAAS

1 INTRODUCTION

An expert system can be regarded as a special category of computer program that seeks to perform expert tasks such as interpreting spectral data. Many remarks and examples in this paper concern infrared spectrometry, but hold equally well for NMR, UV, MS, and other spectroscopic techniques. The interpretation of a spectrum by an expert system should ideally lead to results identical with or even better than those produced by an expert. At the moment expert systems are rather primitive and still in their infancy, but progress is being made rapidly. It is in no way a problem to transform solid rules, like group frequency correlations, into computer programs, though the intensity criteria weak, medium, and strong, ask for a creative translation. Difficulties arise however with the less lucid criteria and concepts that experts implicitly apply. Examples of these criteria will be discussed in the following sections. Spectral features frequently used in the interpretation process are summarized in Table 1.

Table 1 *Spectral features*

* spectral appearance	*band maxima or minima
— region	— transmittance
— presentation	— absorbance
* band intensities	* bandshape(s)
— absolute/relative	— needle sharp
— strongest band	— normal
— number of bands	— broad
— threshold	— shoulder/overlapping bands

Some features, *e.g.* the position of the strongest band or the total number of bands, can be incorporated simply into a program, but serious problems show up with spectral appearance and to some extent also with bandshapes. A lot of creative thinking will have to be undertaken before such criteria can be programmed. Theoretical considerations, spectroscopic as well as chemical, are also taken into account during an interpretation process. Some of the spectroscopic items are presented in Table 2 and they all play a role in the interpretation of a spectrum.

Take the C=C stretching vibration as an example. According to theory this vibration is infrared inactive (no absorption) as long as the group is symmetrically substituted. From the absence of absorption in the C=C stretching region one may conclude that (i) the group is absent, (ii) one or more C=C groups are present, but either symmetrically substituted or the absorption is too weak to be observed. The presence of a single band in the C=C region points to one or more asymmetrically substituted C=C groups with coinciding absorptions. However one or more non-absorbing symmetrically substituted C=C groups may also form part of the molecule. Even the presence of two distinct peaks in that region is ambiguous as they could both arise from one single C=C present in two **optically** different positions. It follows that for theoretical reasons the presence/absence of a band in the C=C region offers a large number of possibilities, which should all be considered carefully during interpretation. Another factor complicating the automation of the interpretation process is the well known fact that the spectrum of a compound depends largely on the sampling technique. For example spectra from the gas phase differ markedly from those of the liquid phase; frequency shifts and intensity changes are observed. Consequently these effects have to be incorporated one way or another in the spectrum–structure correlation. A variety of techniques is currently available as Table 3 shows and an ideally designed expert system should be able to cope with all of them. Other parameters influencing the spectral appearance are the scanning conditions, *i.e.* the instrumental conditions under which a spectrum is run. Examples are presented in Table 4. An expert system should be able to do what an expert does under these circumstances. Inspection of correlation charts reveal that several functional groups absorb in the same spectral region. Discrimination will be possible only if other group-specific evidence is available (other absorption bands or pre-knowledge about the

Table 2   *Theoretical features*

* intensity ($d\mu/dq=0$?)
  — number of each functional group in a molecule
* overtones and combination bands
  — $2\nu_i$ and $\nu_i \pm \nu_j$
* symmetry
  — degeneracy, coinciding bands
* rotational fine structure
  — PQR branches

* configurations, rotamers
  — optically different molecules
* crystal lattice
  — band splitting
* coupling
  — band splitting

Table 3 *Sampling technique*

transmittance, reflectance, emission
temperature, pressure
gas, liquid, solid, solution, matrix isolated
amount of sample, concentration
PAS, ATR, IRRAS, SPECULAR, DRIFT, *etc.*
impurities (*e.g.* CO_2, H_2O, *etc.*)

Table 4 *Scanning conditions*

resolution
accuracy
precision
signal/noise

possible chemical structure). By now, it will be clear that a lot of experience and a good photographic memory, is of utmost importance in the interpretation process. Interpretation is in many ways (still) an art.

2 DESCRIPTION AND COMPARISON OF EXISTING SYSTEMS

Several attempts have been made to develop automated interpretation systems capable of performing expert tasks. The complexity of the human interpretation process has been recognized, however, and therefore a simplified model of the reasoning process has been used. This model can be described as consisting of a number of steps: (i) the translation of spectral data into structural fragments, (ii) the combination of these fragments into complete structures, and (iii) the prediction of theoretical spectra and comparison with the spectrum of the unknown. An important aspect that is not directly a part of the interpretation process is the acquisition and coding of the knowledge necessary to perform interpretations. Although there exist large numbers of correlation charts for many different classes of compounds and it is possible to copy these into computer memory, it would be preferable to extract spectrum–structure correlations automatically from the spectra of a large number of known compounds.

Not all of the systems that will be discussed are capable of performing all of the above-mentioned steps. In Table 5 an overview of some currently existing systems and their application area is shown, whereas Table 6 illustrates their basic capabilities.

2.1 Functional Group Analysis

A general outline of the interpretation module of an expert system is given in Figure 1. One of the most important parts of any expert system is its know-

Table 5   *Overview of the application area of several interpretation systems*

System	IR	MS	^1H–NMR	^{13}C–NMR	UV
DENDRAL[1]	−	+	−	+	−
CHEMICS[2]	+	−	+	+	−
STREC[3]	+	+	+	−	+
CASE[4, 5]	+	+	+	+	−
PAIRS[6]	+	−	−	−	−
SEAC[7]	+	−	+	−	+
EXPERTISE[8]	+	−	−	−	−
ASSIGNER[9]	+	+	−	−	−
EXSPEC[10–12]	+	+	−	−	−

Table 6   *Basic capabilities of several interpretation systems*

System	Automated knowledge acquisition	Automated interpretation	Structure generation	Spectrum simulation	Explanation facility
DENDRAL	+	+	+	+	−
CHEMICS	−	+	+	−	−
STREC	−	+	+	+	−
CASE	+	+	+	−	−
PAIRS	+	+	−	−	+
SEAC	−	+	+	−	−
EXPERTISE	+	+	−	−	−
ASSIGNER	−	+	+	−	−
EXSPEC	+	+	+	−	+

[1] R.K. Lindsay, B.G. Buchanan, E.A. Feigenbaum, and J. Lederberg, 'Applications of Artificial Intelligence for Organic Chemistry: The Dendral Project', McGraw-Hill, New York, 1980.

[2] K. Funatsu, C.A. Del Carpio, and S. Sasaki, *Fresenius' Z. Anal. Chem.*, 1986, **324**, 750.

[3] L.A. Gribov and M.E. Elyashberg, *CRC Crit. Rev. Anal. Chem.*, 1979, 111.

[4] M.E. Munk, M. Farkas, A.H. Lipkis, and B.D. Christie, *Mikrochim. Acta*, 1986, **II**, 199.

[5] B.D. Christie and M.E. Munk, *J. Chem. Inf. Comput. Sci.*, 1988, **28**, 87.

[6] H.B. Woodruff, S.A. Tomellini, and G.M. Smith, in 'Artificial Intelligence Application in Chemistry', ed. T.H. Pierce and B.A. Hohne, ACS Symp. Ser. 306, Washington, 1986, p. 312.

[7] B. Debska, J. Duliban, B. Guzowska-Swider, and Z. Hippe, *Anal. Chim. Acta*, 1981, **133**, 303.

[8] T. Blaffert, *Anal. Chim. Acta*, 1986, **191**, 161.

[9] G. Szalontai, Z. Simon, Z. Csapo, M.Farkas, and G. Pfeifer, *Anal. Chim. Acta*, 1981, **133**, 31.

[10] H.J. Luinge, 'EXSPEC, a Knowledge-Based System for Structure Alanysis of Organic Molecules from Combined Data', thesis, Utrecht, 1989.

[11] H.J. Luinge, G.J. Kleywegt, H.A. van 't Klooster, and J.H. Van der Maas, *J. Chem. Inf. Comput. Sci.*, 1987, **27**, 95.

[12] H.J. Luinge and J.H. van der Maas, *Anal. Chim. Acta*, 1989, **223**, 135.

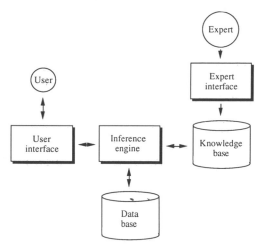

Figure 1 *Schematic depiction of an expert system*

ledge base, a data base containing knowledge about the problem domain to be covered. This knowledge can be stored in a variety of forms of which simple if–then rules (*production rules*) are the most common. The inference engine is a program that decides in which order the rules in the knowledge base are applied to a particular problem. As the knowledge base and the inference engine are well separated, knowledge can be added or modified without altering the overall structure of the system. Hence, expert systems can be developed for different applications by using the same inference engine and changing only the knowledge base. This possibility has led system developers to build so-called *empty shells*, expert systems with an empty knowledge base. Generally, such shells are equipped with advanced user interfaces, explanation facilities, and rule editors. The user interface takes care of the information exchange between the user and the system. Depending on the specific application it can be equipped with graphical facilities or a language processor for problem-oriented communications. An expert system should be equipped with an explanation facility capable of explaining the reasoning process of the system. Usually, this is a program that keeps track of the rules that are applied to a specific problem and that is capable of presenting these rules in a logical order to the user. Absence of such a facility makes the system behave like a black-box and obstructs acceptance of its conclusions. Finally, a rule editor makes it possible to add, modify, or delete knowledge from the knowledge base without having to know details of the programming language used for the construction of the system.

Currently existing expert systems for infrared spectral interpretation simplify the interpretation process as performed by human experts. Interpretations are performed by postulating that a particular structural fragment is present and searching the knowledge base for interpretation rules necessary to accept or reject the fragment. This is called a model driven approach and it is the method used in all systems for the interpretation of vibrational spectra. The

EXSPEC system [10-12] is capable of performing a data-driven approach as well. In this approach a spectral feature is selected from the spectrum and possible correlated structural fragments are listed. Ultimately, the results of both approaches are identical. However, incorporation of a data-driven approach allows the user to tackle a problem from different viewpoints and, hence, increases the user-friendliness of a system.

Interpretation rules used in automated interpretation systems contain spectrometric knowledge describing correlations between spectral and structural features. The types of spectral features used differ from system to system: some only use wavenumber and intensity peak maxima, others also include data on bandshape or on the number of bands within a specified wavenumber range. Furthermore, other types of rules can be added, *e.g.* rules that describe spectral shifts due to different sample states.

Application of the rules to the spectrum of an unknown compound generally results in a list of possible structural features ordered by some kind of certainty factor that is related to the likelihood that the particular feature is present in the unknown. These certainty factors can be assigned to a spectrum–structure correlation by an expert or they can be calculated from large numbers of example data. In some cases complex functions are defined that are used to calculate the probability that a fragment is present depending on the exact position and intensity of a characteristic spectral feature. The CHEMICS system[2] starts with a huge set of simple structural fragments that can occur in a chemical compound. Consecutively, spectral data are used to filter out all fragments that are not supported by spectral evidence. Hence, certainty factors are reduced to 'definitely absent' and 'possibly present'.

2.2   Knowledge Acquisition

When trying to develop an expert system for a particular problem area, one has to consider how to obtain the knowledge that is necessary to solve problems. This knowledge acquisition problem is usually a bottle-neck in the construction of expert systems. Knowledge can be acquired in several ways, all of which involve transferring the expertise needed for high-performance problem-solving in a particular domain from a source to a program. The source is generally a human expert but can also be the empirical data, case studies, or other studies from which a human expert's own knowledge has been acquired. The process of translating the knowledge from the source to the program may be performed by a so-called *knowledge engineer* or by a program.

In the systems previously mentioned, knowledge is acquired in only three different ways: (i) via a knowledge engineer, (ii) via an intelligent editing program, or (iii) an induction program. The 'knowledge engineer approach' is applied in the CHEMICS,[2] STREC,[3] SEAC,[7] and ASSIGNER[9] systems. It has the disadvantage that knowledge has to be acquired from an expert during time-consuming interviews. Furthermore, the knowledge engineer has to have programming experience in order to transform the interpretation rules obtained into computer code. The PAIRS[6] program has been provided with an

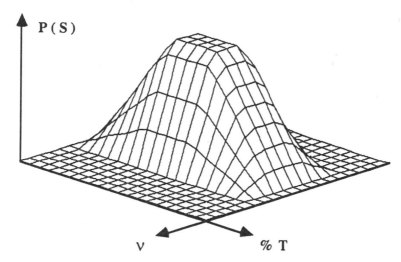

Figure 2 *Probability distribution as used in EXPERTISE and EXSPEC. (P(S) is the probability that a structural fragment S is present, when an absorption at wavenumber ν with an intensity of T% transmittance is present)*

advanced user-interface for entering knowledge. Rules can be expressed in a specially developed language (CONCISE) and are compiled in order to obtain tables that can be used by the program.

As expertise in the field of spectrometry is generally obtained by studying large numbers of spectra during several years, one can assume that it is possible to speed up this process by using computers for the production of spectrum–structure correlations from large numbers of example cases. Attempts to automate the knowledge acquisition process in this way have been described for the CASE,[4] PAIRS,[6] EXPERTISE,[8] and EXSPEC systems.[10,11] In the first two of these programs an occurrence *versus* wavenumber distribution is determined from peak tables of infrared spectra of a specific class of compounds. Characteristic spectral regions for the class can be selected from this distribution and expectation or confidence values are assigned to each region based on the respective occurrence values. Each region can be subdivided into smaller sub-regions for which partial expectation or confidence values are calculated. The result is a class spectrum profile for each class of compounds considered consisting of characteristic regions, confidence values, and any other band descriptions (*e.g.* sharp, broad, weak, and strong). In the EXPERTISE[8] program characteristic spectral regions are determined by finding spectral 'areas' that contain absorptions of all compounds in the specific class. From all regions thus found the ones with the smallest area are selected. Unknowns having absorptions in these intervals have a high probability of belonging to the specific class. Interval boundaries are made less abrupt by applying a linearly decreasing probability distribution as is shown in Figure 2. In EXSPEC a similar approach is used to that in the EXPERTISE program. However, instead of using areas as selection criterion for each spectral region, their information content is calculated.

Although rule generators as described above can be very useful for finding characteristic spectral regions, they yield only part of the knowledge that is generally necessary for solving complex interpretation problems. Information on characteristic spectral patterns, on exceptions to 'established' facts, or on the use of non-spectral data, has to be derived in other ways, *e.g.* by interrogating experts.

2.3   Structure Generation

In many cases one is interested in the entire molecular structure of an unknown compound. Translation of spectral data into structural fragments, therefore, has to be followed by the combination of these fragments into possible structures. In order to achieve this the molecular formula of the unknown compound must be available, except for the EXPERTISE[8] program which is the only system that attempts to generate structures without this knowledge. Obviously, the latter only works for very simple molecules, as it is generally not possible to determine from a spectrum the exact number of each fragment present. Several algorithms have been described that try to solve the problem of generating all possible combinations of fragments without generating duplicates. Several of these (CHEMICS,[2] STREC[3]) have been demonstrated not to be exhaustive,[10] which is a serious problem as it means that the correct solution to a problem might not be found.

The algorithms applied can differ widely. Some generate sets of structural fragments and link these together to form complete molecules. Others use atoms as building blocks and check for consistency with the spectral data during the generation process. Generally, it is possible to use additional constraints, *e.g.* partial structures (possibly determined from other sources) that are prohibited or required in the resulting structures. Recently a structure generator has been described[4] that starts from a hyperstructure containing all possible atoms and bonds. Structures are obtained by removing bonds from this hyperstructure in all possible ways, ultimately resulting in a set of structures in accordance with the data. The method is said to make more efficient use of fragments that have to be present in the resulting structures.

2.4   Spectrum Prediction

Prediction of spectra from structures is only possible with DENDRAL[1] and STREC[3]; DENDRAL concentrates on mass and STREC on vibrational spectra. In the DENDRAL program use is made of knowledge of bonds that are likely to break in a fragmentation process in order to predict a spectrum. The STREC system applies semi-empirical forcefield calculations. This program uses a library containing information concerning geometry, force constants, and electro-optical parameters of molecules and standard fragments. This library is used to formulate the system of equations necessary for calculation of the overall spectra. Examples of computed and observed absorption curves have been obtained for many small molecules. However, it remains to be

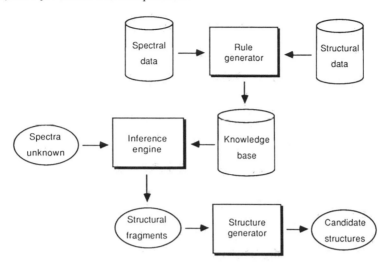

Figure 3 *Schematic depiction of the EXSPEC system. (The squares denote program modules, the ellipses temporary data, and the cylinders permanent data)*

shown if this approach yields similar results when applied to larger, more complex, molecules.

3 THE EXSPEC SYSTEM

The EXSPEC system[10-12] was developed for the interpretation of combined infrared and mass spectral data. The system was written in LPA MacPROLOG and runs on Apple Macintosh™ computers. PROLOG is one of the languages that were especially developed for programming artificial intelligence applications. It is well suited for the representation of spectrum–structure correlations and for manipulating coded chemical structures. In combination with the Macintosh it provided a fast and easy-to-use development environment. A depiction of the system as it currently exists is given in Figure 3. As can be seen from the figure, EXSPEC consists of three parts corresponding with: (i) automated acquisition of knowledge, (ii) application of this knowledge in order to translate spectral features into structural fragments, and (iii) generation of complete structures from these fragments.

Automated knowledge acquisition commences by defining a structural fragment for which spectrum–structure correlations have to be (re)discovered. From a data base of coded structures all compounds containing the specified unit are retrieved. (Partial) structures are coded by explicitly stating the building blocks that are present (currently CH_3, CH_2, CH, C, OH, and O are used as building blocks) and the bonds that connect them (single, double, triple, or aromatic). Next, spectra from a group of compounds containing the structural fragment of interest are selected from a spectral data base and characteristic spectral intervals are determined. For each of these intervals the information content is calculated by applying information theory as described in reference

Table 7   *Output of the rule generating program[11] of the EXSPEC system*

***Rule generation process started using: infrared data

Total number of 5 compounds containing ->CH_x-CO-CH_x<- (cyclopentanone) used.

Total number of 489 compounds without ->CH_x-CO-CH_x<- (cyclopentanone) used.

Maximum information content: 0.082

Interval: [1760, 1744, 93, 89]	Information content: 0.061
Interval: [3505, 3455, 12, 6]	Information content: 0.050
Interval: [1163, 1142, 81, 51]	Information content: 0.038
Interval: [2897, 2880, 64, 15]	Information content: 0.035
Interval: [1409, 1391, 62, 36]	Information content: 0.035
Interval: [977, 959, 51, 27]	Information content: 0.035

Combined intervals:

Interval: [1760, 1744, 93, 89]	Combined information content: 0.061 (75.2%)
Interval: [1163, 1142, 81, 51]	Combined information content: 0.082 (100%)

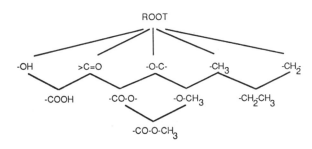

Figure 4   *Part of the structural network used during the interpretation process*

11. At this stage use is made of the entire spectral data base. The information content of each interval depends on: (i) the number of compounds of the class of interest which have an absorption in the particular region, (ii) on the number of interferences (*i.e.* compounds absorbing in the region but not belonging to the class of interest) and (iii) on the total number of spectra considered. Hence, the information content reflects the usefulness of a spectral region for interpretation purposes. Regions are combined into sets for which again the information content is calculated. Combinations with the highest value are used as interpretation rules. An example of the output of EXSPEC's rule generator is shown in Table 7. Interpretation rules obtained in this way are stored in the knowledge base and used for interpreting spectra of unknown compounds.

Spectral data can be entered into the system in three ways: directly from disk, via keyboard, or interactively, *i.e.* the system asks the user for the presence of specific bands in the spectrum. During the actual interpretation use is made of a network of structural fragments as shown in Figure 4. A fragment is selected from this network and the corresponding rules are applied only after the presence of its building blocks has been established. The results of an interpretation are listed in decreasing order of probability of occurrence in the unknown compound. Explanations for each of the conclusions can be obtained in terms of the rules that have been applied during the reasoning process.

When the interpretation has been completed the structural fragments obtained can be combined into complete structures using the structure generator of the EXSPEC system. After having entered the molecular weight of a compound into this program together with any constraints on the elements, rings, or unsaturation present, all possible molecular formulas are generated. Selection of one of these formulas and addition of any constraints that were derived during the interpretation of the spectrum, yields all possible structural formulas.

4 FINAL REMARKS

Automated spectrum interpretation by using expert systems has gained an increasing amount of attention in recent years. Most systems, however, were built as pure research systems running on large mainframes making them difficult for novice users to handle. With EXSPEC an attempt has been made to develop a system that runs on a small computer and that contains several facilities in order to achieve an enhanced user-friendliness.

At the moment expert systems still operate as experienced novices rather than on the level of experts for most problems. This is probably due to the fact that the amount of information that is extracted from a spectrum is still much lower when performed by a computer than by a human. As discussed previously, an expert uses much more information from a spectrum than just peak maxima, intensities, and bandwidths. Incorporation of additional knowledge remains an objective for future research. However, this does not mean that expert systems are not useful at the moment. An expert system can be a good assistant of an expert especially because it exhaustively considers alternative possibilities. Expert systems in structure elucidation should not therefore be regarded as replacements of true experts but merely as new pieces of equipment that can be used to obtain increased performance.

Currently, increased attention is being paid to the computer simulation of neural networks as existing in the human brain.[13] Application of such methods to vibrational data in combination with the possibilities of parallel computing might result in a more effective solution to the structure elucidation problem in the future.

[13] S. Grossberg, 'The Adaptive Brain', II, Amsterdam, 1987.

CHAPTER 4

Data Manipulation in Spectroscopy

A. S. GILBERT

1 INTRODUCTION

Data manipulation deals with the realm of operations lying between successful acquisition of data from the spectrometer and the extraction of information. That is to say those processes that convert raw data into forms more suitable for quantitative analysis, measurement of peak positions and other operations that gather facts as a prelude to gaining physiochemical knowledge about the system under study.

Data acquired from a spectrometer consist of a spectrum of intensity values belonging to channels separated either by equal frequency proportional intervals (wavenumbers) or equal wavelength proportional intervals (angstroms, nanometres, or micrometres). In the simplest category of data manipulation to be discussed the data channels remain independent, thus an output datum in any particular channel is a function of input data in that channel only. Transmittance to absorbance is manipulation of data that belongs to this category. Only one set of data is output, but more than one set of data may be input, for example when subtraction is performed.

A more complex category is represented by the operation of smoothing data to enhance signal-to-noise or sharpening spectral bands to enhance resolution. Here the data channels do not remain independent; any output datum is a function of input data from many channels, but with a heavy bias towards channels that are adjacent to each other and to the output. This is achieved by convolution of the input data set with, *e.g.* a triangular function (in the case of smoothing) to generate one output set of data.

The most complex category to be considered is where one input set of data can be used to generate two or more output sets. The data channels do not remain independent because the data processing involved is not linear, *i.e.* the process cannot be reversed unlike the first and (in formal terms) the second categories described above. This third category of process is iterative and consists of the Maximum Entropy Methodologies (MEM) utilized to improve resolution while simultaneously reducing noise.

There are also a small number of less commonly used manipulations that do not fall into the above classifications and will not be dealt with, *e.g.* Kramers–Kronig analysis.

2   DATA PROCESSING INVOLVING NO MOVEMENT OF DATA BETWEEN CHANNELS

2.1   Transmission to Absorption

Infrared and ultraviolet visible spectrometers measure the transmittance, T which is the fraction of radiation remaining after passage through the sample. This quantity is not proportional to sample concentration or path length but can be converted to one that is, by the relation

$$A = \log_{10} \frac{1}{T}$$

where A is the absorbance.

Most if not all modern UV–visible spectrometers do this automatically but it is normal for the operators of FTIR spectrometers to initiate this conversion themselves in software.

The statement of the linear relationship of absorbance to sample parameters such as concentration is called Beer's Law (though it should perhaps be renamed after Bouguer and Bernard[1]). However, apparent deviations from this law may occur to a greater or lesser extent. These are usually due to physiochemical effects, which will not concern us further, but can also arise from settings of instrumental parameters. For instance where the actual width of a band is of the order of or less than the spectrometer's slit function, both the band height and area will be modified by differing and non-proportional extents depending on the absorption strength. This is because the transmission profile (or shape) of the spectrum varies with sample concentration and path length due to its logarithmic relationship with the absorbance and will therefore be convoluted (see below) in a different way with the slit function. Varying the slit function from experiment to experiment introduces even more deviations (NB the effective slit function in an FTIR spectrometer may be a combination both of the 'physical' resolution (maximum path difference) and a software apodization function implemented transparently to the user).

Apart from these problems with Beer's Law, conversion to absorbance of spectral regions of low transmittance will lead to large roughly exponential increases in noise compared to regions of higher transmittance. This is princi-

[1] D.T. Burns, in 'Advances in Standards and Methodology in Spectrophotometry', ed. C. Burgess and K.D. Mielenz, 1987, p. 1–19.

pally due to the noise being either virtually independent of transmittance (IR) or proportional to $\sqrt{T}$ (UV–visible), but can also come from lack of dynamic range during digitization.

These and other considerations normally limit accurate quantitation using IR absorbance bands to about 1.5 au, as long as spectral subtraction is not employed when the limit can be somewhat lower, but it is possible to rely on UV–visible bands of up to 3 au or so providing they are at least an order of magnitude broader than the instrumental slit width and the instrument has a low stray light specification.

Measured Raman spectral intensity is, like absorbance, proportional to sample concentration.

2.2 Spectral Subtraction

Raman and most FTIR spectrometers are single beam devices and, as vibrational spectra of solvents are commonly rich in bands, spectral subtraction routines are a central part of any associated data system. Most manufacturers of commercial systems implement these routines in a variety of ways, the user having the choice of doing a single subtraction with fixed constants, or doing the process interactively whereby the subtraction constant is varied up or down in small amounts until a satisfactory end result is achieved. Many data systems also possess an 'automated' subtraction mode where if the user is confident that a particular spectral region contains only solvent bands he can instruct the computer to adjust the subtraction constant until the region is flattened, with the hope that complete 'cancellation' of all solvent bands across the spectrum will occur.

Spectral subtraction is performed on absorbance data for IR and UV–visible spectroscopy. Removal of interfering solvent bands is probably the most common application, but removal of 'impurity' bands using a 'pure impurity' reference spectrum as subtractant is probably just as popular. Well prepared alkali-halide discs often yield sufficiently good concentration proportional IR spectra so that spectral subtraction can be carried out with a fair degree of confidence. This enables quality control to be performed on solid samples by subtraction of the reference spectrum of a known *pure* sample from, say, a production batch of the same compound to look for possible impurities. Such a process will not always work for solids as the method of preparation of discs and the alkali halide matrix itself can induce sample-to-sample spectral variation even if polymorphism does not occur. Other techniques such as diffuse reflectance and photoacoustic absorption may overcome some of these difficulties.

But as with all types of data manipulation, care must be exercised in case the conditions are inappropriate for the process. Subtraction always increases the noise across the spectrum and increases can be very dramatic in regions where highly absorbing bands are present. These are precisely the circumstances where Beer's Law deviations become significant and such conditions can easily lead to the generation of artifacts with the danger of them being mistaken as bands. The accuracy of qualitative analysis by IR is reduced for absorbance values greater that 1 au. Greater absorbances can be tolerated for UV–visible absorption and strong Raman bands generally cause fewer problems because their associated signal dependent photon statistical noise increases as the square root of the signal strength. But Raman spectra can, like IR, suffer from band shape changes brought about by physiochemical interactions and it should be noted that Raman detectors (both PM tubes and arrays) can suffer from saturation so that very strong bands are depressed in apparent intensity.

2.3 Baseline Correction

Ideally the spectral trace should not possess any deviations from the horizontal other than those from the bands of interest, *i.e.* infrared or electronic absorptions in the case of IR and UV–visible spectra and vibrational scattering in the case of Raman spectra. However, the latter are often underlain by broad fluorescence emission lowering signal/noise and solid phase IR spectra are commonly affected by scattering which reduces transmission towards higher frequencies, again lowering S/N and also restricting dynamic range. These problems can only be properly corrected by attending to the sample, but data manipulation is appropriate where the effects are not too severe and where display of the spectrum would result in the peaks of interest being small and 'squashed'. It is of course necessary if peak positions are to be accurately measured where they lie on sloping baselines, but care should be exercised that broad bands are not assigned to fluctuations in the background. This is particularly important for IR spectra from highly scattering alkali-halide discs where broad absorptions, for example from [→NH]⁺ stretching, can be lost in the rapidly falling transmission curve.

Baseline correction is performed by ratioing transmission data or subtracting from absorption or Raman spectra an artificially constructed background. While it might be expected that the background of transmission spectra from solid state samples should follow some simple or known scattering function, the distribution of particle size and shapes causing the effect is too complex, though parabolic curves are sometimes a reasonable approximation. Most commercial software usually offers both parabolic and linear functions which can be applied in an automatic mode but bearing in mind the caveat in the previous paragraph, it is better for the operator to use his own judgement and construct a baseline of his own by selecting suitable points throughout the spectrum and 'joining' them up. Unfortunately, some commercial software is crude and will only 'draw' straight lines between the operator selections; instead polynominal interpolation should be implemented so as to give a smooth curve.

3   LINEAR DATA PROCESSING INCLUDING DATA TRANSFERS BETWEEN CHANNELS

Differentiation and smoothing are well known data manipulations of this category, the former is extensively used to resolve or separate bands in UV–visible spectroscopy, the latter is commonly applied to IR and Raman spectra on account of the greater noise manifested in these techniques. Savitzky and Golay[2] have developed convolution functions for both these operations; these functions are essentially filters which are passed over the spectrum, the convolution operation being described below.

$$A' = \sum_{x=-m}^{x=m} F_x A_{n+x}/F$$

In fact any roughly triangular (convex) function can be used to smooth data, but Savitzky–Golay functions are designed to minimize distortion to space in most circumstances. [Examples are illustrated in Figures 1(a) and (b)].

The outcome of smoothing is to reduce the noise on the data, therefore allowing better estimations of peak positions, *etc.*, but this is always achieved at the expense of reducing band height and increasing band width (thus reducing resolution). The amount by which these parameters change depends on the relation of the width of the smoothing function to the spectral bandwidths, thus the relative band heights of differently shaped bands in a spectrum change, but fortunately the areas under bands are unaffected so that there is still a means of quantification available. This is important if the noise characteristics of a set of spectra being compared require different degrees of smoothing. The optimum trade-off between reduction in resolution or noise is achieved by smoothing with a function that exactly matches the line shape (matched filter), but this is not, of course, possible if the band shapes differ.

Differentiation always increases the noise in a data set, and has therefore chiefly been employed on UV–visible absorption spectra with their typically high signal/noise characteristics. By selecting various of the Savitzky–Golay functions [*e.g.* Figure 1(c)] an attempt can be made to find a suitable combination of smoothing and order of differentiation. Even order differentiation does

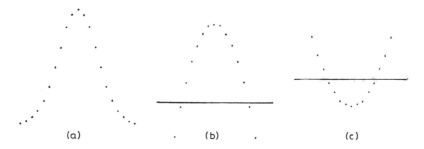

(a) (b) (c)

Figure 1 *Convolution functions: (a) 21 point Gaussian smoothing function* (m=10); *(b) 15 point Savitzky– Golay smoothing function* (m=7); *(c) 15 point Savitsky–Golay second differential function* (m=7)

[2] A. Savitzky and M. Golay, *Anal. Chem.*, 1964, **36**, (8), 1627.

produce band-like profiles, but relative band area relationships are lost and side lobes are generated. Broad bands with their gentler slopes tend to be suppressed in relation to sharp bands, but analytes often yield narrow bands whereas other components in mixtures or solvents often give broad sloped backgrounds so that differentiation offers some advantages in these situations.

Both smoothing and differentiation are part of the general process of convolution. This is best considered by examining the data in the Fourier domain. Figure 2 shows the Fourier transforms (only the cosine coefficients are shown in this and succeeding figures) of Lorentzian bands. The waves decay exponentially from the origin, the broader the spectral band the more rapid the decay. It is helpful to think of the Fourier domain as a frequency domain (though this designation is arbitrary); sharp spectral bands therefore yield rapid oscillations with many more high frequency components (waves) in the Fourier domain (data appearing much further from the origin) whereas broad spectral bands give waves concentrated much nearer the origin (rapid decay). According to the convolution theorem, convolution in one domain is equivalent to a point-by-point multiplication in the other domain, the multiplying function being the Fourier transform of the convoluting function.

The effect of multiplying the Fourier transform data of a Lorentzian doublet [Figure 3(a)] by a smoothing function is to cause the waves to decay more rapidly so that the resultant bands in the spectral domain are broader [Figure 3(b)]. The effect of differentiation is to multiply the contents of the Fourier domain by a rising function, illustrated in Figure 3(c) for the case of 2nd order differentiation, and this naturally brings about a sharpening of the resultant spectral bands.

A more elegant method of sharpening bands is to match the convolution function to the band shape. This is the basis of the so-called Fourier Resolution Enhancement (FRE) or Fourier Self Deconvolution (FSD) methods. The real spectrum can be looked upon as consisting only of spikes which are convoluted with the line-shape to yield observed bands of finite width and the task is therefore to *de*-convolute these bands by dividing in the Fourier domain with the Fourier transform of the line-shape function. This is the same as multiplying (*i.e.* convoluting) with the reciprocal of the function. The coefficients of the Lorentzians decay by $\exp(-f)$ so that multiplication by the function $\exp(f)$ gives a set of pure cosine waves which when transformed back to the spectral domain yield an infinitely sharp line (Figure 4).

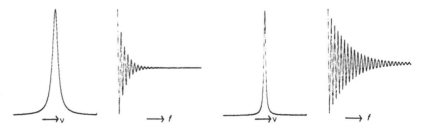

Figure 2 *Lorentzian bands and their Fourier transforms*

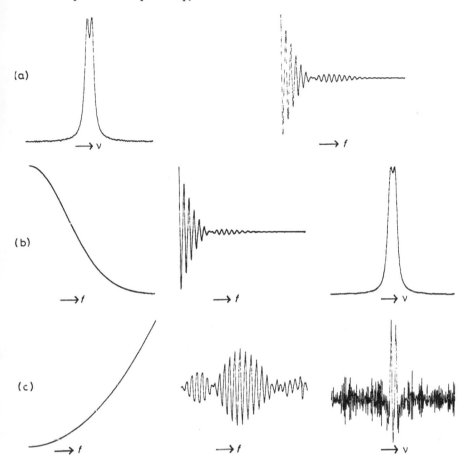

Figure 3 *The effect of smoothing and differentiation on a spectrum of overlapped Lorentzians: (a) Lorentzian doublet and its Fourier transform; (b) Fourier domain multiplying function for Gaussian smoothing, the resultant Fourier domain after multiplication of this with the Fourier transform (a) and the final smoothed spectrum (obtained by reverse Fourier transformation of the resultant Fourier domain; (c) Fourier domain multiplying function for second order differentiation, the resultant Fourier domain after multiplication of this with the Fourier transform from (a) and the final sharpened spectrum obtained as in (b). (NB Inverse of normal second differential is obtained)*

Sadly such an ultimate degree of sharpening cannot be achieved in practice because of noise which is ever present in data, if not from the detector then by virtue of the finite word length of the computer used to contain it. This noise will usually be roughly equally distributed across the Fourier domain, multiplication by $\exp(f)$ will vastly increase the higher frequency components and because noise in any one channel in the Fourier domain appears in *all* channels in the spectral domain, the resultant effect is that the spectral bands will be swamped. Fortunately the problem can be partially alleviated by apodization, that is using a second function to reverse the exponential rise of

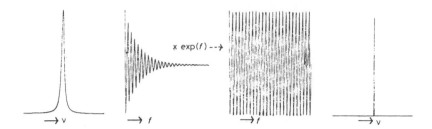

Figure 4 *Complete deconvolution of Lorentzian line-shape from a noiseless Lorentzian band (unrealizable in practice)*

the first in the high frequency region. There is a penalty however in that the apodization function imposes its own shape on the final band. This is illustrated in the case of Gaussian apodization in Figure 5. The net result is that a useful degree of resolution enhancement is obtained with control of noise. This is brought about because the overall 'weighting' of the combined multiplication function causes the Fourier coefficients (waves) of the spectral bands (signal) to decay less rapidly in regions of the Fourier domain where they are significant (*i.e.* middle-range frequency components are enhanced) whereas the high frequency region where noise predominates is suppressed. The weighting function can therefore be thought of as a frequency band pass filter, whereas smoothing and differentiation are respectively high and low cut-off frequency filters.

Both the decree of convolution and apodization can be separately controlled using FRE (FSD). Weakening the apodization gives narrower bands, but at the expense of higher noise [Figure 6(a)]. Weak deconvolution leaves some of the original lineshape and thus breadth in the bands [Figure 6(b)], but over deconvolution generates negative side lobes [Figure 6(c)].

The combination of Lorentzian Deconvolution Gaussian Apodization

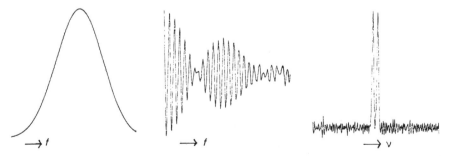

Figure 5 *Complete deconvolution with apodization, data used being the Lorentzian doublet shown in Figure 3 (a). Illustrated is a Fourier domain multiplying function which is the product of a rising exponential and a Gaussian smoothing function, the resultant Fourier domain after multiplication of this with the Fourier transform from Figure 3(a) and the final spectrum (obtained by reverse Fourier transformation of the Fourier domain)*

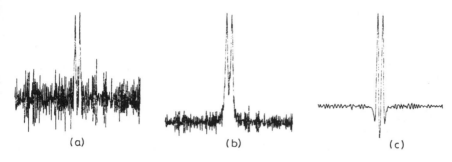

Figure 6 *Effects of varying the degrees of deconvolution and apodization (LDGA function): on the Lorentzian doublet from Figure 3(a). (a) Weak apodization; (b) Weak deconvolution; (c) Over-deconvolution by using a rising exponential appropriate to a broader Lorentzian band*

(LDGA) has been extensively utilized[3] for the resolution enhancement of proton NMR spectra for many years and is particularly suited to this technique because of the generally consistent and fairly pure Lorentzian line-shape of the observed bands. The information gained are splitting patterns and coupling constants that would not be revealed without recourse to much higher magnetic fields. The treatment of vibrational or electronic spectra is usually more difficult because of the more varied line-shape of the bands. Even where the line-shapes are known (the nearly Lorentzian line-shapes of many IR bands in liquid/solid phases) they are usually very variable in width. Complete deconvolution of broad bands therefore result in over-deconvolution of narrower ones, with consequent generation of side lobes. However a lower degree of deconvolution can sometimes be used to separate the sharp bands and they can then be removed by peak stripping before proceeding to tackle the broader ones.

The trade-offs between bandwidth reduction, R and noise increase, NI are given below for the LDGA weighting function which demonstrates that apodization only controls the noise within limits.

R	NI (final:original spectrum)
1	0.63
0.8	1.06
0.6	2.85
0.4	38.8
0.3	1270
0.2	7×10^6

(NB– Figures are for a completely deconvoluted Lorentzian being converted to a pure Gaussian. They will vary slightly depending on the exact width of the band)

[3] J.C. Lindon and A.G. Ferrige, *J. Mag. Res.*, 1981, **44**, 556.

Alternative apodization functions can be used[4] which do not allow the noise to expand so quickly for a given reduction factor, but this is at the expense of broader wings to the final bands and anyway, whatever type of apodization used, noise increases in an exponential manner for R factors below about 0.25.

Gaussian bands may also be deconvoluted but require sharper apodization to counteract the power two exponential deconvolution function. The degree of narrowing possible is smaller than for Lorentzian bands because the noise is harder to control.

As long as the weighting function varies smoothly from a value of 1 at the origin, then the area under bands remains constant, whether they are subjected to broadening or narrowing. If side-lobes are generated (from over deconvolution) the area under negative lobes subtracts from the positive regions to equate with the original band area. This area constancy is a consequence of the fact that the value of the interferogram at the origin is equal to the integrated area under the spectrum.

Figure 7 shows original and LDGA enhanced IR spectra of a sample of Meningococcal Serogroup C polysaccharide after treatment with a carbodimide. The analytical problem was to quantitate the amount of internal lactone and a bi-product represented by carbonyl bands at 1745 cm^{-1} and 1695 cm^{-1} respectively. While the separation of these bands is not perfect their total strengths can at least be very much better estimated than from the original spectrum.

The treatment of the IR C-H stretching region of a long chain hydrocarbon containing molecule is shown in Figure 8 demonstrating separation of the methyl and methylene bands.

The LDGA enhancement of part of the UV spectrum of *para*-cresol is shown in Figure 9. However, UV–visible spectra are in general not so amenable to line analysis since they possess varied band-shapes so that it is usually not worthwhile to use specific line-shapes for deconvolution. Simple but suitable frequency band pass filters can be constructed for use in the spectral domain by the convolution process and a recent publication[5] demonstrates their general superiority over differentiation on the resolution of UV spectra for purposes of quantitative analysis.

4 MAXIMUM ENTROPY METHODOLOGIES (MEM)

The convolution methodologies described in the above section suffer from a number of limitations. Firstly they only discriminate between signal and noise on the basis of frequency, thus noise in the same frequency band as signal cannot be rejected without also losing information. Secondly, they are unconstrained so that unreal (*e.g.* negative) outcomes can ensue, and thirdly, there is no quantitative measure by which the rightness of the result can be judged. These problems can however be addressed by the Maximum Entropy Method.

[4] J. Kauppinen, D.J. Moffat, H.H. Mantsch, and D.G. Cameron, *Appl. Spectrosc.*, 1981, **35**, 271.
[5] R. Jones, *Analyst*, 1987. **112**. 1495.

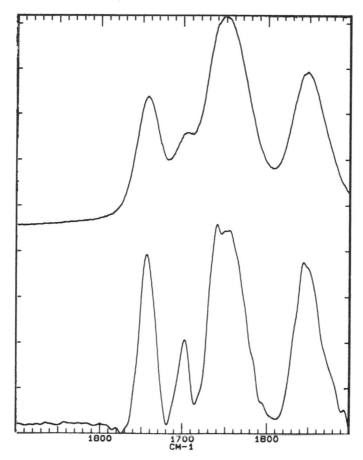

Figure 7 *FRE of an IR spectrum (LDGA function). (top) Original; (bottom) final*

Consider the real spectrum, in astronomers parlance the object that it is desired to obtain; astronomers were the first people to use Maximum Entropy Methodology to resolve optical problems. This may be broadened by molecular collisions or uncertainty processes and will be viewed through a 'slit' of finite width to give a 'blurred' image. In addition this image will be degraded by noise resulting from both the statistical (photonic) nature of the signal and the thermal background of the detector (usually one or the other predominates). In simple (de-)convolution operations, use is made only of the operator's knowledge concerning the 'blurring' function but the operator also has information about the noise, *i.e* the precision that can be assigned to each data value in. the image.

There will exist a large number of other spectra that have the same total intensity as the image. Individual intensity values are all either positive or zero. These spectra can be separated into two classes. One class, that of impossibles (non-candidates) contain spectra that when blurred by slits, *etc*. and then

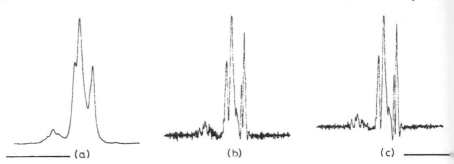

Figure 8   *LDGA resolution enhancement of the* C–H *stretching IR bands of a hydrocarbon chain containing molecule. (a) Original spectrum; (b), (c) final spectra produced by using two slightly different degrees of deconvolution*

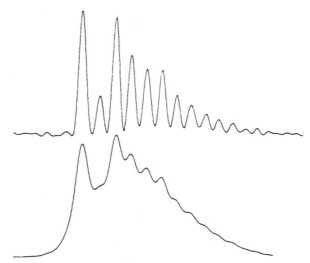

Figure 9   *LDGA resolution enhancement of part of the UV spectrum of* para-cresol. *(top) Final; (bottom) original*

subtracted from the image, give residuals that are different from the known noise. The other class, that of the possibles contain spectra that when similarly blurred and then subtracted from the image yield residuals that cannot be distinguished from the known noise (*i.e.* they have the same variance). The object is obviously a member of the possible class and the problem is now to distinguish which one it is. The object can never be known with complete certainty, because only the overall properties of the noise are known, not its actual value in each data channel. Therefore the choice of which candidate to assign as the object must be made on the basis of likelihood, *i.e.* the candidate possessing the maximum likelihood is considered to be the object.

The first person to derive an expression that quantified spectral likelihood was Frieden.[6] If the intensity values in each data channel are taken to be a (finite) number of photons then the likelihood of any given spectrum can be expressed as the number of ways, W that the total number of photons, N can be arranged among the channels to give that spectrum.

$$W = \frac{N!}{(N-n_1)!n_1!} \times \frac{(N-n_1)!}{(N-n_1-n_2)!n_2!} \times \ldots \ldots \ldots etc.$$

$$= \frac{N!}{\prod_{v=1}^{v=M} (n_v)!}$$

where n_v = number of photons in each channel v
and M = number of channels.

The larger W (*i.e* the greater its degeneracy), the more likely is that spectrum. For very large N (approximating a continuum) Stirling's approximation can be applied and putting $P_v = n_v/N$ gives

$$Log_n W = \text{Constant} - \sum_{v=1}^{v=M} P_v \log_n P_v = S$$

The logarithmic nature of the expression causes it to be termed the entropy, S of the spectrum and the task is thus to find which of the possibles possesses the largest value of S (*i.e* $\log_n W$), in other words the candidate with the greatest or maximum entropy. This is not however simple as in general the number of candidates is *very* large.

It can be tackled however by convoluting a trial spectrum with the blurring function and subtracting the result from the image. The residuals are used to modify the initial trial spectrum to give a new spectrum which is again convoluted and new residuals found and so on. The modifications to each new spectrum must be done in such a way that S is always maximized, any negative values are set to zero and the process, which should begin with a straight line trial spectrum, continues until the variance of the residuals is reduced to that of the known noise on the image. The trial spectrum at that point is then considered to be the object. The modifications to the trial spectra at each cycle of the process are determined by employing the differential forms of S and the residuals and quite simple schemes can be constructed but, while useful in some cases often suffer from instability and normalization problems that derive from the particular form of S described above.

A different philosophy to that of Frieden has been adopted by Skilling and co-workers[7] to derive an alternative form for the entropy which does not generate problems of normalization.

[6] B.R. Frieden, *J. Opt. Soc. Am.*, 1972, **62**, 511.
[7] S.F. Gull and J. Skilling, *IEE Proc. (F)*, 1984, **131**, 646.

$$S = \sum_{\nu=1}^{\nu=M} (a_\nu - a_\nu \log_n a_\nu/c) - Mc$$

where c = constant

The rationale is that the maximum entropy spectrum must be that candidate that embodies no extra correlation or structure other than those imposed by the image, *i.e.* it is the simplest (flattest or smoothest) of the candidates.

Both this form of S and the variance of the residuals are convex functions in the multi-dimensional space defined by the data channels. Convex surfaces touch only at one point, thus there is a unique spectrum (the object) that possesses the maximum entropy for a residuals variance equal to the noise on the image. Skilling[8] has also developed a robust algorithm that performs a highly efficient search for the object and all the results described below have been obtained using it.

Figures 10(a), (b) show the MEM results from the Lorentzian doublet shown in Figure 3(a) and the spectrum in Figure 8(a) respectively. The same Lorentzian lineshapes utilized for FRE (Figure 5, 8) were employed as the blurring functions, but it is obvious that MEM has yielded a considerably greater degree of resolution enhancement compared to the simple deconvolution methodology. In addition there is no observable noise on the MEM spectra; strictly there should be no noise at all as the object is by definition noiseless, but where very noisy images are utilized the objects given by MEM often show some. This could result from providing the algorithm with an incorrect value for the noise, but where this has been defined correctly the reason must be sought elsewhere. The noise on the image may not have Gaussian statistics (the algorithm can be modified to take account of other types of noise); alterna-

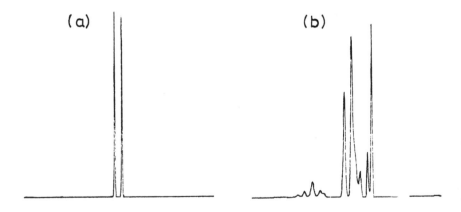

Figure 10   *MEM resolution enhancement (a) The Lorentzian doublet from Figure 3(a) (b) The IR spectrum from Figure 8(a)*

[8] J. Skilling and R.K. Bryan, *Mon. Not. R. Astron. Soc.*, 1984, **211**, 111.

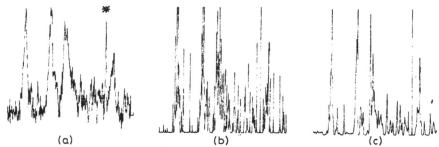

Figure 11 *MEM resolution enhancement of the Raman spectrum of bovine serum albumin in* H_2O. *(a) original spectrum (b), (c) final spectra produced by using two different values for the noise. The blurring function was taken as Lorentzian with linewidth equal to the asterisked peak.*

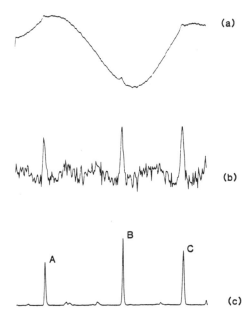

Figure 12 *Baseline correction and resolution enhancement by MEM. (a) Original spectrum; weak bands on top of a broad fluctuating background; (b) Result of conventional baseline correction only; (c) MEM treatment as above and described in text.*
This figure courtesy of Dr. J. Skilling, St. John's College, Cambridge.

tively the object given by MEM may not be the true object (MEM only finds the most likely candidate).

Figures 11(a)–(c) show the result of deliberately varying the value for the variance of the noise in a Raman spectrum of bovine serum albumin. By allowing too low a value to be used [Figure 11(b)] certain noise 'peaks' are treated as signal and appear in the object. By using too high a value to give Figure 11(c), the noise has been very effectively suppressed, but at the possible expense of certain small 'real' bands from the sample.

Figure 12 illustrates the result of separating a Raman spectrum into two parts on the basis of spectral band width. Here two objects, a broad background from fluoresence and a much sharper vibrational spectrum, are simultaneously sought for by the algorithm with the constraint that both together must fit the image and its noise after convolution with the blurring function. This effectively generates a spectrum automatically corrected for baseline.

Finally it should be noted that MEM deconvolution can, like simple FRE, suffer from incorrect specification of the blurring (line-shape) function. However instead of generating negative side lobes from over-deconvolution the MEM algorithm utilized here usually fails to converge to a final result (*i.e* no object can be found that will give residuals as low as the noise variance). This is a reflection of the fact that the algorithm has been given an impossible task.

5 SUMMARY

MEM enables simultaneous improvement in spectral resolution with reduction in noise unlike the simple deconvolution methods (FRE) which almost always suffer trade-off between these properties. Over and above specification of line-shape (blurring) function required by both techniques, no apodization is necessary, and the only further assumption needed for MEM is the noise on the image. The result of MEM deconvolution/resolution enhancement (the object) is always positive, unreal negative spectra are not allowed by the theory. These advantages are, however, gained at the expense of computer time though this is by no means excessive. The results shown here were all obtained in a few minutes using a PDP 11 computer.

CHAPTER 5

Spectroscopy and Chemometrics

B. DAVIES

1 INTRODUCTION

Molecular spectroscopy has always been recognized as a rapid and flexible analytical technique for obtaining information from materials. The type of information needed and the methods used to extract the information from spectroscopic data have both changed radically in the last few years and reflect the current trends in the use of analytical science.

Whatever else may have changed the fundamentals of spectroscopy have not. The position and shape of an absorption feature on the wavelength (or frequency) scale is indicative of the material chemistry and its interactions with its environment or matrix and the intensity of the feature is related to the amount of material present in the sample under investigation. Both these measures may be modified by a variety of physical and chemical effects but they and their baselines are the primary source of spectral information on the sample. Whatever technique is used to convert these spectral data into relevant information involves extracting it from these types of features.

Traditionally the spectroscopist's eyes, linked often to years of experience have been used to look at sets of spectra, filter out unnecessary clutter, and extract information from the data. This is then put into the context of the application or problem under investigation and reported to the spectroscopist's customer.

The move in modern spectroscopy has been from subjective to objective investigations of spectra and consequently from manual to computer aided or automated assessment of data. This approach has been fuelled from two sources.

The use of analytical science in many industries is now changing from laboratory based techniques to those matched to investigating and controlling production processes. Here the application is matching analytical technologies to the business needs of increased production control, higher quality products, and better use of resources. It is the shift from using the analytical information in a reactive way to one of a proactive nature. Spectroscopy, in particular UV, NIR, and FTIR, is well suited to this type of application. The techniques are rapid, can use minimal sample preparation, and are robust enough in both an analytical and mechanical sense to fit into production areas. In these applications the spectrometers are effectively sensors for processes. Their role is to gather high quality validated data and pass this on for conversion to relevant process information.

The changes in the use of spectroscopy have come at the same time, and have been aided by a revolution in data processing techniques. Mathematical, statistical, database, and expert system techniques have come together to be known as chemometrics. Chemometrics is a powerful set of tools for investigating the information content of data sets. Many of these tools have been used in a variety of spectroscopic application areas.

This chapter will cover the practical principles involved in using some of these chemometric tools on spectral data, highlighting their use in pharmaceutical applications. Emphasis is given to the techniques of experimental design, sample selection, discriminant analysis and cluster analysis which are frequently overlooked in spectroscopic applications.

2   SPECTRA AND MULTIVARIATE METHODS

Investigating the information content of spectra is an exercise in multivariate data analysis. The information content of spectroscopic measurements is potentially very high; qualitative and quantitative chemical information is usually present and, depending on the technique, physical information may also be available. This information is spread throughout the wavelength range of the spectrum and each wavelength element or digitized data point in a spectrum will contain some of this information and may be considered as a separate variable.

In a typical NIR experiment up to 100 spectra, each containing 350 wavelength points can represent such a data set. The scale of the data becomes greater when one moves into the mid-IR range where a single spectrum may contain over 3500 data points. For such applications it is necessary to use techniques which can reach into potentially very large data sets and isolate areas or patterns in the data representing the information content. The essential patterns in the data can then be presented to the user in a simple graphical form and used to build a working mathematical model of the chemical system under investigation. This model can then be applied to a spectrum of a test sample and used to predict the property under investigation.

A considerable number of multivariate data analysis methods exist and hardly a month goes by without the literature reporting some new algorithm or an extension of a mature approach to a new problem. To the user this presents a problem as to which approach to use on their spectral data.

Chemometric methods fall into essentially two categories; classification or discriminant methods and multivariate calibration techniques. The approach to using them is similar and well worth considering before a more detailed look at the techniques themselves.

3 CALIBRATION AND TRAINING DATA SETS

Chemometric methods require a calibration or training set of data to build a model of the chemical system under investigation. In the spectroscopic application a set of materials must be created or obtained which can be used to generate the spectroscopic data fed into the modelling package. This set of materials must contain sufficient chemical, physical, or other property information about the system under investigation for a robust model to be generated. In quantitative analysis applications the samples must represent the working range of concentrations of the components to be assayed. They must also be chosen to bring out any interaction effects between components as these may have profound spectroscopic effects. If the spectroscopic technique in use can be affected by matrix effects then the samples should contain this information.

In summary the learning set of samples used to train the chemometric technique must contain all of the sources of chemical and matrix variances likely to be seen in test samples to be assessed. If this is the case a good chemometric technique can then construct a model of the system which is robust and capable of providing both correct and reliable predictions of the sample properties of interest. A model cannot be expected to handle effects in test samples of which it has no prior knowledge, except perhaps to warn the user that the sample has un-explained and un-modelled sources of variance.

It follows from this that some considerable thought must be put into just what types of materials and how many of them must be used as a chemometric training set. If, in a quantitative analysis application, there are three components to be assayed, in a solid material of variable particle size which also has a variable moisture content then the training set must contain enough spectroscopic information for all five of these properties to be modelled. The user is presented with having to prepare or obtain samples with a representative spread of all of these properties and their potential interactions—a daunting task! Using traditional approaches to preparing samples for calibrations this could lead to many tens of samples being prepared and even then not all combinations or levels of components being obtained. To maximize the information content and minimize the preparation effort Experimental Design techniques are often used to suggest an experimental scheme for the calibration set. These techniques are now readily available in textbooks, computer programs, and well supported by training courses. An understanding of their use in developing calibration sets is vital to the analytical spectroscopist, and an integral part of the experimental system, as valuable in the creation of a robust method as choosing the correct spectroscopic technique or chemometric algorithm.

4 EXPERIMENTAL DESIGN

A useful experimental design for the three component system above is seen in Figure 1. This is a central composite design, and is shown in its coded form. The levels 0, −1, 1, −1.68179, and 1.68179 are scaling values to be applied to the users own concentration values. The 0 refers to the mean level of the property of interest and the −1, 1, −1.68179, and 1.68179 are scaled to span the property space of interest. In the example of a quality control assay the 0 level could represent the nominal concentration of a component, the −1 and 1 levels the top and bottom of the specification. The assay would then be well calibrated with respect to the specification, for example 90% and 110% of the nominal concentrations, the 1.68179 and −1.68179 coding levels are from the geometry of the design, 1.68179 is $(2(2^{0.5}))^{0.5}$, and give information on the interaction effects. The central composite design is a hybrid of a star design and a factorial design and represents an experimentally efficient method. Experimental designs to suit individual applications are readily available in both coded tabular form and as computer programs.

Often the spectroscopist is attempting to model real world systems and the calibration issues can become very complex. Samples prepared in the laboratory, even those produced in pilot plants cannot always contain the same variances as those seen in production samples. It is very difficult, or impossible to duplicate fully, industrial processing in the laboratory. This forces the user out of the experimental design approach into the arena of choosing calibration samples from a population of real samples. This problem is especially seen in many NIR applications on natural products; an experimentally designed set of wheat samples would be impossible to produce. If experimentally designed laboratory samples are impossible to produce or would not contain enough process information to be useful the spectroscopist must turn to the 'real' samples themselves as a source of calibration.

Choosing which samples are to be used and then obtaining their assay or property values is a difficult task. If samples are coming from an established process they may all be very close to the nominal specification values of the assay properties of interest. Any calibration obtained from such a set would be very limited and potentially not robust with respect to the full specification range of the material. In many cases a 'better' calibration can be obtained from samples taken from a process that is out of control than one that is fully controlled!

Process or natural product samples must be assayed by some other analytical technique; quite often it is this technique that the spectroscopic assay is replacing. These assay values are then used with the sample spectra to calibrate the spectroscopic technique. The spectroscopic assay is then firmly linked with the secondary technique, in effect, through the model that emulates it. If the secondary technique has a poor accuracy or precision then the spectroscopic technique may be compromised. If this approach is used then the user should be very critical of the secondary calibration technique used.

(a)

Component 1	Component 2	Component 3
−1	−1	−1
−1	1	1
1	−1	1
1	1	−1
0	0	0
0	0	0
−1	−1	1
−1	1	−1
1	−1	−1
1	1	1
0	0	0
0	0	0
1.68179	0	0
−1.68179	0	0
0	1.68179	0
0	−1.68179	0
0	0	1.68179
0	0	−1.68179
0	0	0
0	0	0

(b)

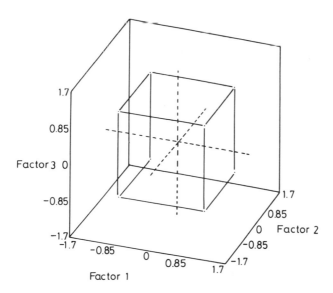

Figure 1 *An experimental matrix for 3 component system using a central composite design: (a) Coding matrix; (b) Plot of the design matrix*

There are several strategies for choosing which samples should be used from those available.

If the samples are already being assayed by some other technique then the user should choose samples which best represent the calibration space of interest and best approach the codings an experimental design would give. For applications with several components the sample assay values are best analysed by Principal Component or Factor Analysis and samples chosen from plots of the latent variables.

If the samples are not already being analysed then it is often impractical or expensive to analyse all of them, especially as only a fraction may contain the necessary calibration information for the spectroscopic assay. Here the spectra themselves can be used to provide information about which samples should be used for calibration. The spectral data can be analysed using Principle Component or Factor Analysis and the resultant latent variable plots used to chose samples which best represent the variances in the spectral data, it is these variances which the modelling technique will use. These samples can then be analysed by some other technique to give the assay values for the properties of interest. The samples should now represent the best source of information to the modelling technique of the variances associated with the 'real world', but they may represent a very limited calibration range. Sometimes a hybrid calibration set can be established. Here 'real world' materials are mixed with laboratory produced samples from an experimental design. The laboratory samples give the assay range information and the 'real' samples provide the production variances and matrix effect information. Often this approach will not work if the differences between synthetic and 'real' samples are too great; the modelling technique may evaluate them as two quite separate clusters of spectra and be unable to produce a robust model.

5   MODELLING THE SPECTROSCOPIC DATA

Once the calibration set of spectra has been obtained it is then analysed with a chemometric technique and a model of the system created. It should be appreciated by all users that a model is fitted to the data and not the data fitted to a model. Invariably a model, if flexible and complex enough can be fitted to any spectral data and care must be taken not to overfit. The ultimate test of a model is not how it will behave with the calibration set of spectra but how well it predicts the properties of interest from a set of spectra it has never seen, the validation set.

6   DISCRIMINANT AND PATTERN RECOGNITION TECHNIQUES

For qualitative applications chemometric techniques of discriminant analysis, cluster analysis, and pattern recognition are employed. Here sets of spectra representing different types of materials are distinguished from each other. This may be the classical library search, raw material identification, or material quality management application. The differences between the spectra of the materials may represent chemical, physical, processing, or sample preparation differences between the materials.

The basic approach is that of supervised discriminant analysis, where a set of spectra of materials of known chemical or chemical/physical compositions is used as a training set. The technique then processes the data looking for features or patterns in the data which can be developed into a computational decision rule for evaluating similarity and dissimilarity between the groups. This rule is then applied to assign unknown or test spectra.

The second approach is that of unsupervised discriminant analysis. Here a set of spectra are presented to the chemometric algorithm with no prior assumptions made about the number of separate groups present or their relationships. The technique then analyses the data, again looking for features representing similarities and differences between the spectra. Once this process is complete the resulting clusters of spectra can be examined to see if they yield any information on the chemistry, processing or other properties of the material used. This approach can be used to investigate learning sets for quantitative analysis applications. The pattern recognition tools can also be used to look for clustering of calibration data, perhaps into laboratory and production clusters, or look for spectral outliers.

6.1 Supervised Discriminant Analysis—The Mahalanobis Metric

Objectively measuring the similarities and differences between spectra is notoriously difficult. One technique for confidently measuring spectral differences is the use of the Mahalanobis Metric. This approach has been applied successfully to the investigation of incoming raw materials using NIR spectroscopy. It has significant benefits over peak matching and other simple library search systems often used for material identity tests. The technique can be 'taught' and calibrated to respond to the batch to batch or sample to sample variations found in many materials. This allows it to be used to identify materials which may have a broad specification and consequently quite variable spectra. The algorithm is based on simple and sound statistical foundations and results can be given meaningful confidence limits or probabilities.

The application of Mahalanobis Metric calculations to spectroscopic data can best be described with a simple example of a three group discrimination and using only two data dimensions. Figure 2 shows the idealized NIR spectra of three different materials. In this trivial example the spectra can be distinguished by eye, each material having a characteristic response. In the real application there may be 50 or 100 different materials and 10 or more spectra for each material, making visual interpretation impossible or tedious. The three spectra can be distinguished from each other using the relative absorbances at two wavelengths in the spectrum.

Figure 3 shows a plot of the absorbances at wavelength 1 *versus* wavelength 2 for the three spectra. The three materials are now represented by points in this two dimensional plot, the three points being well separated. The distances between the points are a measure of the discriminating ability of the two wave-

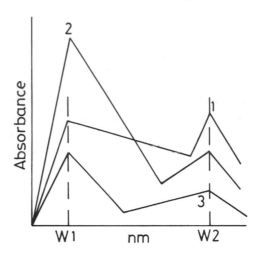

Figure 2 *Mahalanobis Metric: idealized NIR spectra*

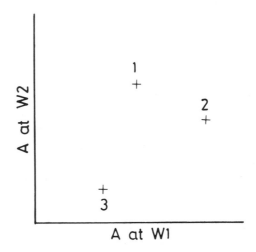

Figure 3 *Mahalanobis Metric: absorbance plot W1* versus *W2*

lengths chosen for the plot. For real materials the use of a single spectrum to define the position of a group is obviously naive. Sample to sample or batch to batch variations will be present in real samples and manifest themselves as variances in their NIR spectra. The algorithm must be 'taught' where these effects are and what are their magnitudes. A much more realistic view leads to multiple samples of each material type being scanned and plotted on the diagram. This leads to the plot seen in Figure 4. Each material group is now represented by a cluster of points around some mean position. The size and shape of this cluster is some measure of that material's batch to batch variation and preparation to preparation variance.

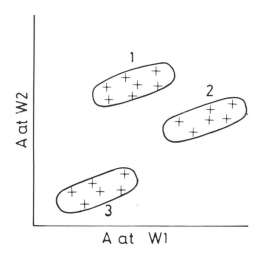

Figure 4   *Mahalanobis Metric: multiple batches*

The circumscribing ellipse around each group in Figure 4 represents the one standard deviation contour of the group dispersion around its mean position. This development of a group, a point swarm around the mean position and associated standard deviation contour, is pivotal to the effective use of the Mahalanobis Metric in predicting to which group an unknown or test material may belong.

Test samples can now be classified by their distance from individual group mean positions. The shorter the distance the higher the probability that the test samples belongs to that group. As the groups have both a shape and direction then the use of a simple Euclidean Metric is not optimum. The shape and sizes of the groups seen in Figure 4 show a definite orientation in the plot. Each group is generally extended along the approximately 45 degree line as a result of particle size differences in the original materials. The concept developed by Mahalanobis is that of a unit distance measure which varies with the direction of measurement. The unit is the 'Mahalanobis distance'. The unit is such that its equivalent Euclidean distance is large in those directions in which a group is elongated. The distance vector is estimated by calculating an ellipse which circumscribes the group data at a distance of one standard deviation of absorbance from the mean position of the group. The measurement metric is defined as the distance from the group mean position to the perimeter of the ellipse along the direction of measurement. The metric is therefore highly directional and reflects both the group shape and direction in the plotted space.

A group boundary can be considered to be at a distance of three standard deviations from the mean position, this approximates to a 99% confidence interval. An unknown material is tested against a library by calculating the Mahalanobis distance of its position to that of the mean positions of the groups in the library. If the distance to a group is evaluated as less than three units then the test material can be reliably assigned to that group, assuming that the distances to the other groups in the library evaluate as greater than three.

Figure 5 shows the plotted position of an unknown sample. It has an equal Euclidean distance to all three groups. When the distances are evaluated with respect to each group standard deviation contour then it is closest to Group 2 and lies more than three units from the other groups. The unknown material would be assigned to Group 2.

In 'real world' applications often up to 10 or more wavelengths are needed to successfully separate the groups in a library, and algorithms which optimize the intergroup distances with a minimum of wavelengths have been developed. If a set of wavelengths cannot be found that reliably separates all the entries in a library from each other then some ambiguity in the assignment of test materials can occur.

In practice the use of Mahalanobis distances as a library search technique is backed up by more incisive full spectral comparisons of test and candidate spectra. This is necessary as in this example only two wavelength values were used to evaluate distances and it is conceivable that important differences between two materials could contribute to their spectra at wavelengths not evaluated in the Mahalanobis calculations. It nevertheless represents an extremely rapid, robust, and statistically reliable means of searching large libraries.

6.2 Unsupervised Discriminant Analysis

Unsupervised discriminant analysis or cluster analysis is complicated by the fact that the technique must not only be capable of creating a discriminant

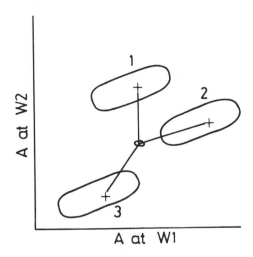

Figure 5 *Mahalanobis Metric: prediction of material*

function, but it must also devise a scheme or strategy for allocating spectra to classes or clusters of 'similar' types.

One method of creating clusters of 'like' spectra is by generating a metric matrix and linking 'like' spectra together. The matrix can be generated by calculating the 'distance' of each spectrum from the others. The spectra producing the shortest distances are linked, then averaged, and the process continues until all spectra are linked. The results can then be interpreted as a dendrogram. This simple approach to linking observations, in this case spectra, into groups is a well established technique in cluster analysis. It is one of the few approaches not requiring some matrix inversion.

A simple example of the use of this approach is seen in Figure 6. The NIR spectra of six materials are shown. It can be easily seen by eye that there are three different types of spectra here;

A, B, and C are similar;

D and E are similar;

F is a lone spectrum differing considerably in its features from the others. The metric used to calculate their 'distances' was:

$$d = \sum_n (A_n - B_n)^2$$

A_n and B_n are the absorbance values of spectrum A and B at wavelength n.

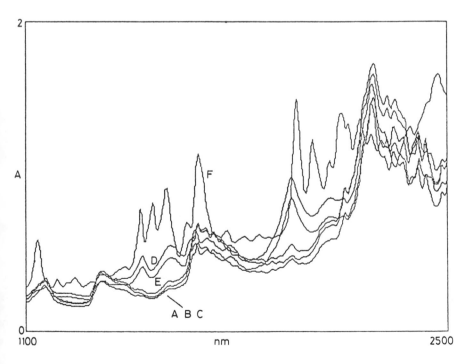

Figure 6 *NIR spectra of six materials*

The magnitude of d is a measure of the similarity of the two spectra. Similar spectra will have low d values, dissimilar spectra high values.

Figure 7(i) shows the metric matrix of the first pass of the spectra through the distance calculations. Spectra B and C are identified as the first pair. Their spectra are then averaged to produce Group B* and the matrix recalculated, Figure 7(ii). Spectra D and E are now identified as the closest pair, their spectra are averaged and the matrix recalculated, Figure 7(iii). Spectra A and B* are now linked. At this point the spectra A, B, and C have been identified as a group, as have D and E, but the spectrum F is still well isolated. If the process is continued then A, B, C, D, and E would be brought together. In this case the separation of the spectra into groups could easily be performed by eye, but the principles are readily applied to many hundreds of spectra and the distance metric changed to optimize chemical or physical property differences between the materials used.

Once the clusters have been established it is then the responsibility of the spectroscopist to give them some chemical or process validity before they can be used as a diagnostic tool.

7   QUANTITATIVE TECHNIQUES

Perhaps the most well developed of chemometric techniques used on spectra are those used to develop multicomponent quantitative analysis applications. Spectroscopy, in particular UV–Visible and mid-IR has always had a role to play in analysing mixtures of materials. In recent years NIR has developed into a tool for routine assay work and by their nature NIR spectra require numerical rather than visual assessment.

The main tools for developing multicomponent assay are those based around the techniques of Principal Component Analysis (PCA) and Partial Least Squares Analysis (PLS). The aim with both these tools is to develop a meaningful mathematical model of the chemical system and use it to predict the properties of test samples from their spectra.

7.1   Principal Component Analysis (PCA)

Spectroscopic data are highly multivariate and often colinear. The purpose of PCA is to give an overview of the dominant information patterns in the data. These are the relationships between the spectra and the relationships between the wavelengths or frequencies, they can then be explored mathematically or graphically. PCA sets out to express the main information of a calibration set of spectra in terms of a lower number of variables, these are the Principal Components of the original data matrix. This has the effect of maximizing the amount of information in a greatly reduced number of variables and allows a much more efficient understanding and manipulation of the data.

It is achieved by performing an eigen analysis of the covariance matrix of the spectral data matrix. The eigenvalues each represent a fraction of the total variance in the data and their corresponding eigenvectors provide the coeffi-

Figure 7(i)

Pass 1

	A	B	C	D	E	F
A	0					
B	4.537	0				
C	5.318	0.381	0			
D	11.947	6.341	9.099	0		
E	4.951	6.851	9.665	3.264	0	
F	48.514	33.667	38.771	19.066	30.668	0

Figure 7(ii)

Pass 2

	A	B*	D	E	F
A	0				
B*	4.832	0			
D	11.947	7.625	0		
E	4.951	8.163	3.264	0	
F	48.514	36.124	19.066	30.668	0

Figure 7(iii)

Pass 3

	A	B*	D*	F
A	0			
B*	4.832	0		
D*	7.633	7.078	0	
F	48.514	36.124	24.051	0

Figure 7 *Distance matrices for cluster analysis*

cients for the Principal Components and the directions of maximum variation through the data. If sufficient eigenvalues are calculated then all the variance in the data will be accounted for, but no data reduction will have taken place. One of the most important aspects of PCA is choosing how many of the components are significant for modelling the original data. This significance is of both a statistical and chemical nature. Indicators exist which measure the statistical significance of each factor and can indicate a cut off point after which adding more factors only models noise, but it is up to the user to test the chemical significance of this cut off. In essence the information content of a large spectral data set is distilled into a small number of Principal Components.

For quantitative analysis applications these Principal Components can then be regressed with the calibration values of the standards to produce the system model. Principal Components which correlate strongly with the calibration set property values are then used to develop a model of the system, those with no or low correlations can be discarded. This model is then used to predict the assay values of a test sample from its spectrum. Together with the assay figures these models can indicate the errors associated with the prediction and the validity of applying the model against the test sample. These are important diagnostics when the approach is part of an automated analysis system and must be made to fail-safe on the assay.

7.2   Partial Least Squares (PLS)

Partial Least Squares (PLS) is now being used as a data reduction/modelling technique for spectral data. Here the approach is similar to that of PCA, extracting a series of Principal Components from the data. PLS differs from PCA in that both the spectral data and the property or assay data are used together in an iterative fashion to build a model. The chemical data are used to find a pattern in the spectroscopic data that correlates with them. This ensures that the estimated regression factors have relevance towards the chemical values.

The benefits of this over PCA are that PLS should be able to model a system with both fewer and more relevant dimensions than PCA, hence the model should be more robust and easier to visualize by the user. Often, in practise PCA and PLS give very similar prediction errors on unknown sample spectra, but in general PLS models have fewer dimensions and are easier to interpret.

8   SUMMARY

Spectroscopy using chemometric data evaluation techniques is now very well established.

Chemometrics are seen as a supplement to the skills of the spectroscopist rather than a replacement.

They allow complex spectroscopic and chemical systems to be investigated and interpreted by greatly simplifying the amount and type of data presented to the user.

The would be user of these techniques has access to algorithms for both qualitative and quantitative applications, choosing the correct one for the application needs some thought.

Chemometrics should not be used in isolation from good experimental design and instruments which produce high quality data, it is only a part of the whole analytical system.

The use of spectroscopy in process areas and the types of problems experienced with process samples will fuel further developments in the chemometrics arsenal.

9 BIBLIOGRAPHY

Experimental Design
S. N. Deming and S. L. Morgan, 'Experimental Design: A Chemometric Approach', Elsevier, Amsterdam, 1987.

Discriminant Analysis
B. F. J. Manly, 'Multivariate Statistical Methods: A Primer', Chapman and Hall, London 1986.
D. Luc Massart and L. Kaufman, 'The Interpretation Of Analytical Chemical Data By Use of Cluster Analysis', John Wiley and Sons, Chichester, 1983.
D Luc Massart *et al.*, 'Chemometrics: A Textbook', Elsevier, Amsterdam, 1988.
D. S. Frankel, *Anal. Chem.*, 1984, **56**, 1011.

Quantitative Techniques
H. Martens and T. Naes, 'Multivariate Calibration', John Wiley and Sons, Chichester 1989.
'Chemometrics, Mathematics, and Statistics in Chemistry', ed. B. R. Kowalski, Reidel Publishing Company, Dordrecht, 1983.
M. A. Sharaf *et al.*, 'Chemometrics', John Wiley and Sons, Chichester 1986.
'Analytical Applications of Spectroscopy', ed. C. S. Creaser and A. M. C. Davies, The Royal Society of Chemistry, London, 1988.

CHAPTER 6

Spectroscopic Detectors for HPLC

B. J. CLARK and A. F. FELL

1 INTRODUCTION

In liquid chromatography there has been a continual search for more detection selectivity as well as sensitivity from the assay method. In the absence of a suitable stable and sensitive universal detector a number of different possibilities have been explored. These include the combination of different modes of detection within the same instrument, with one particular example combining UV, fluorescence, and conductivity (Figure 1).[1]

Another detection format is to tailor the detector to a particular property of the solutes such as in chirally active compounds where polarimetric detection has recently become commercially available. Although these detectors offer considerable flexibility, sensitivity and stability can be problems which have still to be overcome.

An alternative method to address selectivity and sensitivity problems is to combine spectrometry with HPLC in so-called 'hybrid systems'. This is a continually expanding area of current interest with numerous elegant interfacing formats providing the link between the chromatograph and the techniques of NMR, MS, AA, and FTIR.

However in all these cases we are dealing with systems in which problems of compatibility are regularly involved. In NMR a limitation is the difficulty in observing weak analyte signals in protonated solvents. One solution was reported by Laude and Wilkin[2] where they circumvented the problem by using protonated solvents in a flowing system and spin decoupled to remove the solvent effect. Similar problems exist in infrared and a novel method has been developed to interface FTIR to a reversed phase HPLC system[3] where

[1] R.P.W. Scott, in 'Liquid Chromatography Detectors', Elsevier, Amsterdam, 1986.
[2] D.A. Laude and C.L. Wilkins, *Trends Anal. Chem.*, 1986, 5, 60.

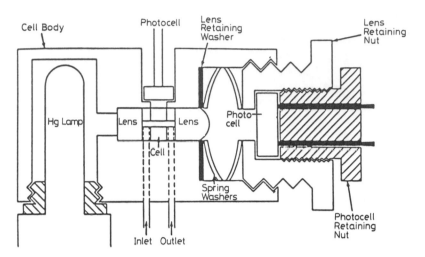

Figure 1   *Diagram of a trifunctional detector which incorporates UV, fluorescence, and conductivity functions within the detector. Alongside the lenses two stainless steel discs constitute the electrodes of the conductivity cell*
(Reprinted with permission from R. P. W. Scott, 'Liquid Chromatography Detectors', Elsevier, Amsterdam, 1986.)

aqueous mobile phases were used in the chromatographic mode and later removed by solute extraction in an aqueous/organic segmented stream (Figure 2).[3] However, although these combined modes of detection have been usefully illustrated in the reported applications, commercial instrumentation has been slow in following this lead.

This is not however the case with HPLC–MS where a number of interfaces exist between the chromatographic column and the mass spectrometer which are successful commercially, albeit at a considerable cost. Of these the two most popular are the Thermospray and moving belt, with the former currently being favoured[4] over moving belt systems. In addition to the use of these interfacing formats the field of LC–MS has been extended further, recently by the commercial combination of MS–MS with HPLC[5] where each MS operates in a different mode. With regard to the chromatographic aspects, recent developments have focused on the advances which include the coupling of supercritical fluid chromatography to MS[6] where, due to the greater compatibility, interfacing can be less of a problem.

In hybrid systems, however the major impact has involved the multichannel spectroscopic HPLC detector with particular emphasis centred on its use in biomedical and pharmaceutical analysis.[7] In commercial terms multichannel

[3] J.W. Hellgeth and L.T. Taylor, *Anal. Chem.*, 1987, **59**, 295.
[4] D. Barcelo, *LC–GC*, 1988, **6**, 324.
[5] H. Poppe, *Chromatographia*, 1987, **24**, 25.
[6] D.E. Games, A.J. Berry, I.C. Mylchreest, and S. Pleasance, *Anal. Proc.*, 1987, 371.
[7] A.F. Fell, B.J. Clark, and H.P. Scott, *J. Pharm. Biomed. Anal.*, 1983, **1**, 557.

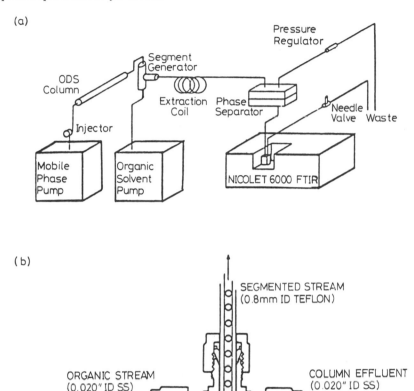

Figure 2 *(a) Diagram of an adapted LC–FTIR system used by Hellgeth and Taylor; (b) Blow-up of the aqueous organic segmented stream generator for solute extraction from the aqueous based reversed-phase HPLC eluent*
(Reprinted with permission from J.W. Hellgeth and L.T. Taylor, *Anal. Chem.*, 1987, **59**, 295.)

detection, based almost exclusively on the linear photodiode array detector (LDA) and predominantly as an absorptiometric detector (although luminescence detectors are available), has been very successful since the first unit was launched in 1981. At present there are about 20 different formats of this detector available from many of the major instrument manufacturers.

These devices permit the rapid acquisition of spectra (many times per second), within a defined wavelength range, during the elution. The basis of the detectors is an array of photodiodes ranging from 35–1024 individual light sensitive units (up to 4096 diodes have also been reported). These are located in the focal plane of a diffraction grating polychromator in a reversed optics configuration (Figure 3). Each photodiode corresponds to a particular wave-

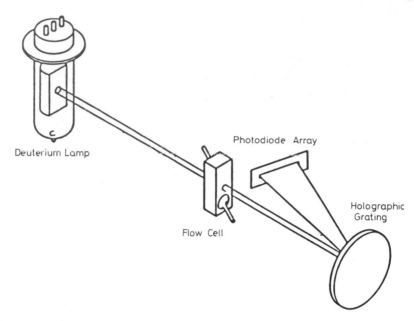

Deuterium Lamp

Photodiode Array

Holographic
Grating

Flow Cell

Figure 3 *Schematic of the reversed optics of the linear photodiode array detector*

length (or narrow wavelength range) where the resolution is determined by the geometry of the polychromator–LDA combination. In the first commercial system introduced in 1981 a 256 array was used to cover the range 190–600 nm. This gave spectral resolution of ±2nm (around 50 diodes were used for reference purposes). It can be considered that the limit imposed on resolution by this detection mode (conventional spectrophotometers 0.25nm) is a drawback, as it could lead to spectral problems with compounds which exhibit narrow spectral bandwidths. As a result loss of the fine structure might be observed and this might lead to measurement errors in absolute absorbance. However, in most cases, and in particular with drug compounds, spectral bandwidths (measured at half-height) are generally large and an instrumental bandwidth of ±2nm is adequate.

Since the first commercial systems, advances in the instrumentation have revolved around the method of scanning the array of diodes in order to obtain the spectral information. In the original format as introduced in 1981 the array was self-scanned which could result in low signal-to-noise (SNR) values due to the noise generated in the integrated circuit multiplexers. One alternative to this was to use an externally scanned detector array as in Figure 4 and to reduce the number of elements in the array, which allowed increased diode surface area with longer integration times[8] and slower scan speeds which

[8] A.G. Wright, A.F. Fell, and J.C. Berridge, *Chromatographia*, 1987, **24**, 533.

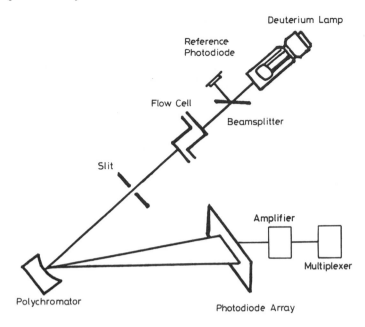

Figure 4 *Externally scanned linear photodiode array detection system, where the signal from the diodes is amplified then taken through a multiplexer. This is in contrast to the self-scanned array where the multiplexer is in front of the amplifier and the noise on this unit is also amplified*

together can result in increased SNR. However to obtain a similar spectral coverage the spectral bandwidth must be increased. Nevertheless these two distinct formats continue to be marketed. In the self-scanned array large spectral ranges are possible with the resultant generation of a large data matrix. In the external scanned system improved SNR is possible, in conjunction with improvements in detection sensitivity, but this mode has the drawback of reduced number of channels and software algorithms are required to interpolate between the spectral data points. Following on from the presence of these different detection formats two distinct instrumental setups have emerged:

(i) in which *complete systems* spectrochromatographic, absorbance, wavelength, and time (A, λ, t) data capture is possible, through the use of computer control and data manipulation of the spectral information from a large array;

(ii) *stand-alone systems* which contain an inbuilt microprocessor capable of collecting a limited number of spectra from a much reduced detector array.

In the industrial analytical laboratories both formats, which differ considerably in cost, have often been utilized, but in separate areas. In the former case the larger data matrix has been best exploited in a research and development environment whereas the stand-alone system has most usefully served the quality assurance area.

2 COMPUTER AIDED METHODS

With both instrumental formats, large amounts of data are generated from each chromatographic run and much of the interest at present is in the computer aided development of suitable digital algorithms to allow data reduction, manipulation, and presentation of the spectral data matrix. The methods developed can be classified as;
- (a) optimization of the chromatographic separation,
- (b) data presentation algorithms,
- (c) methods for solute recognition,
- (d) methods for enhancing detection sensitivity,
- (e) techniques for testing chromatographic peak homogeneity, and purity,
- (f) expert systems.

2.1 Optimization of the Chromatographic Separation

In the first stages of method development in HPLC, parameter optimization is an extremely important factor in achieving adequate separation for a sample mixture. In order to attain this the many variables of the chromatographic system have to be taken into account. With regard to these variables, column selection and the physical parameters of flow rate and temperature all have a very large bearing on the separation. Once a choice has been made in this direction then a primary concern is the choice of mobile phase constituents. Frequently a trial and error or, at best, a univariate approach is adopted in optimizing the mobile phase constituents, but it is becoming apparent to the analyst that a more structured approach should be carried out especially when more than a binary mixture of mobile phase constituents is likely. Basically two modes of structured procedures exist, either simultaneous or sequential methods. These differ in the method of arriving at the final goal. In the former method a predefined scheme is used, whereas in the latter search algorithms are used at each step and direct the next experiment according to the separation achieved by the previous. This latter approach is particularly suited to the LDA format where during the optimization strategy, peak tracking and peak purity assessment can be facilitated by using the spectrochromatographic information contained in the data matrix. Peak tracking through recognition algorithms, by either peak area measurements or by the extent of elution through total integral values, have been shown to be usefully carried out on the LDA.[8]

As regards purity assessments, then a selection of the methods below which can quickly be applied, can be incorporated into an optimization scheme.[9]

[9] A.G. Wright, J.C. Berridge, and A.F. Fell, *Analyst*, 1989, 114, 53.

2.2   Data Presentation Algorithms

The three dimensional data matrix (A, λ, t) representing spectra (A, λ) acquired at sequential time intervals (t), can be readily accessed by computer aided methods to generate the elution profile at any wavelength (λ) and the spectrum at any time (t). The spectrochromatogram (data matrix) can be conveniently presented as an isometric plot of (A, λ, t) at any convenient viewing angle (Figure 5a and b) or as its equivalent two-dimensional 'contour plot' (Figure 5c), where isoabsorptive lines are plotted in the (λ, t) plane.[10] Unfortunately, although these presentations provide some help in establishing the optimum detection wavelengths and for indicating the presence of components in a mixture, they have little additional value.

2.3   Methods for Solute Recognition

Although spectral libraries are common in mass spectrometry and in infrared they are not as frequently used in HPLC. A number of reasons exist for this and include the often low information content of the UV spectra and its susceptibility to solvent effects which in HPLC include; pH, type of organic mobile phase modifier, and temperature.

Archiving of data for spectral recognition should however be an advantage of the LDA detector if these problems can be overcome. Spectral libraries were first reported in LDA by Fell and co-workers,[11] where an 'inverse file' structure was used together with archive files of normalized spectra (Figure 6). Library methods are now established and have been reported using both spectral and retention data[12, 13] and following on from this, a number of the instrument companies have introduced commercial software packages.

2.4   Methods for Enhancing Detection Sensitivity

When first released in the marketplace it was considered that the LDA was 'down' on detection sensitivity over conventional variable wavelength detectors. However, as was shown in the work by Fell[11] the detection sensitivity was in the same order of magnitude as the conventional systems.

Moreover it has been illustrated that sensitivity can be enhanced through improvement in SNR by combining adjacent photodiodes in the array. A global method of carrying this out is to sum the outputs from all the photodiodes (where the whole of the data matrix is presented in 2-dimensions) to give the total absorbance chromatogram as in Figure 7 (200–400nm summed).[14] This is analogous to the total ion current technique in mass spectrometry and can be used to indicate the presence of trace impurities in the chromatographic profile[12] in a rapid preliminary search.

[10] A.F. Fell and B.J. Clark, *Eur. Chromatogr. News*, 1987, 1, 16.
[11] A.F. Fell, B.J. Clark, and H.P. Scott, *J. Chromatogr.*, 1984, **316**, 423.
[12] D.M. Demorest, J.C. Fetzer, I.S. Lurie, S.M. Carr, and K.B. Chatson, *LC-GC*, 1987, 5, 128.
[13] D.W. Hill and K.J. Langner, *J. Liq. Chromatogr.*, 1987, **10**, 377.
[14] B.J. Clark, A.F. Fell, H.P. Scott, and D. Westerlund, *J. Chromatogr.*, 1984, **286**, 261.

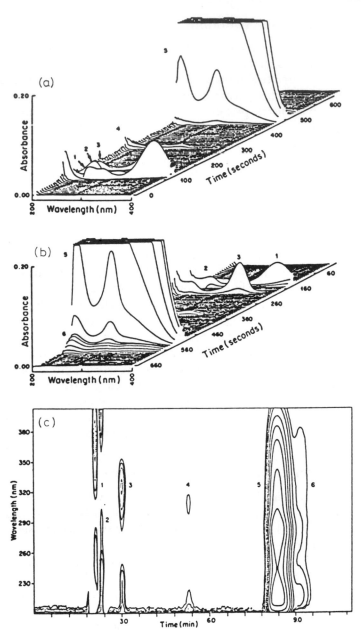

Figure 5 *Pseudoisometric spectrochromatograms: (a) forward; (b) reverse projections; (c) contour presentations for azathioprine and related impurities. Generated with a photodiode array detector (HP1040A) coupled to a HP85 microcomputer and separation by reversed-phase ion-pair HPLC*

(Adapted from A.F. Fell and B.J. Clark, *Eur. Chromatogr. News*, 1987, **1**, 16.)

   Key to peaks: (1) *1-methyl-4-nitro-5-thioimidazole*; (2) *1-methyl-4-nitro-5-thioimidazole*;
   (3) *6-mercaptopurine*; (4) *1-methyl-4-nitro-5-chloroimidazole*; (5) *azathioprine*; (6) *synthetic process impurity*

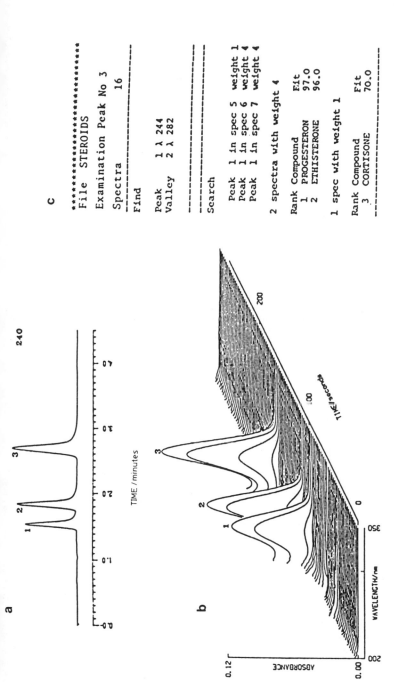

c

```
*************************
* File  STEROIDS
*
* Examination Peak No 3
*
* Spectra        16
*
  Find
*
* Peak     1 λ 244
* Valley   2 λ 282
*
----------------------------
  Search
*
* Peak 1 in spec 5    weight 1
* Peak 1 in spec 6    weight 4
* Peak 1 in spec 7    weight 4
*
----------------------------
  2 spectra with weight 4
*
* Rank Compound         Fit
*   1  PROGESTERON      97.0
*   2  ETHISTERONE      96.0
*
----------------------------
  1 spec with weight 1
*
* Rank Compound         Fit
*   3  CORTISONE        70.0
*
----------------------------
```

Figure 6 *The implementation of archive retrieval algorithms in multichannel HPLC–UV detection for the identification of closely spectrally similar steroids in a resolved chromatogram: (a) conventional chromatogram at 240 nm; (b) the respective isometric projection, illustrating the close spectral similarity; (c) library search for peak 3 from 16 archive spectra, to give a short list, based on comparison of inverse files on the maxima and minima to which a weighting is assigned and a best fit is calculated*

(Adapted with permission from A.F. Fell, B.J. Clark, and H.P. Scott, *J. Chromatogr.*, 1984, 316, 423.)

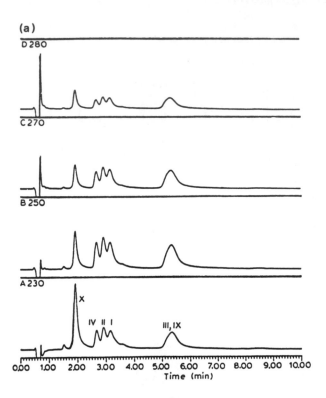

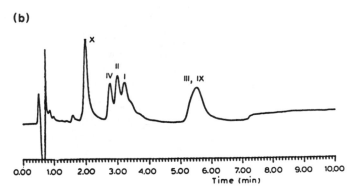

Figure 7   *Multiple-wavelength chromatograms for zimeldine (I); norzimeldine (II); zimeldine-N-oxide (III); 3(4-bromophenyl)-3-(3-pyridyl)allylamine (IV); 3-(4-bromophenyl)-N-hydroxy-N-methyl-3-(3-pyridyl)allylamine (IX); E-isomer of II as internal reference standard (X).*

      (a) *obtained by cutting through the 3-dimensional data matrix of (A, λ, t); (b) total absorbance chromatogram from 200 to 400 nm*

(Adapted with permission from B.J. Clark, A.F. Fell, H.P. Scott, and D. Westerlund, *J. Chromatogr.*, 1984, **286**, 261.)

As stated above sensitivity can be enhanced by a more limited use of the 'diode bunching' method by combining the output from adjacent photo-diodes and this can result in an improvement in SNR.[15] This effect is however to be expected when one considers that the noise aspects of a detector can be decreased in proportion to the square root of the number of datum points averaged. In addition it has been shown that SNR improvements by ensemble averaging can be obtained not only in the wavelength domain but also in the time domain. In diode bunching for the optimum improvement in detection sensitivity a relationship has been reported between the detector bandwidth (DBW) and spectral bandwidth (SBW). As an example in Figure 8 the steroid ethynyl estradiol gives a mean value of the ratio DBW:SBW which was found to be 0.42 for an LKB Model 2140, LDA dectector (LKB, Bromma, Sweden) which gives a four fold improvement in SNR[11] when a bandwidth of 10 nm is used above the default condition of ±2nm. For other commercial detectors a slightly different optimum ratio exists. This exercise can also be carried out in

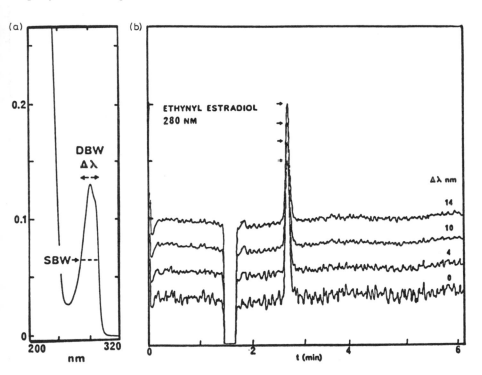

Figure 8 *Effects of increasing the detection bandwidth (DBW), (diode bunching) on ethynyl estradiol: (a) optimum improvement in signal-to-noise ratio is achieved when the DBW (in this case spectral bandwidth at half-height is 24 nm) is equivalent to about 0.4 of the spectral bandwidth; (b) reduction in the noise level with diode bunching up to 10 nm at 280 nm (default value as start point, equivalent to 4 nm (but coded to 0 in the diagram). Above 10 nm both the signal and the noise level decrease*
(Adapted from A.F. Fell, B.J. Clark, and H.P. Scott, *J. Chromatogr.*, 1984, **316**, 423.)

[15] E.M. Kirk, B.J. Clark, and A.F. Fell, *Chromatographia*, 1987, **24**, 759.

the time domain as shown for the steroids; progesterone, ethisterone, and cortisone acetate where a factor of two in enhancement is achieved (Figure 9).[11]

These methods have principally been reported for the detectors which have the capability of collecting a data matrix of (A, λ, t). On the other instruments with limited detector arrays, developments have involved detection sensitivity enhancements through longer pathlength flow cells, larger single detector elements which allow longer integration times, and the aforementioned externally scanned detector arrays. As a result these developments have given quite impressive increases in detection sensitivities with the resultant improvements in the quality of the spectral information.

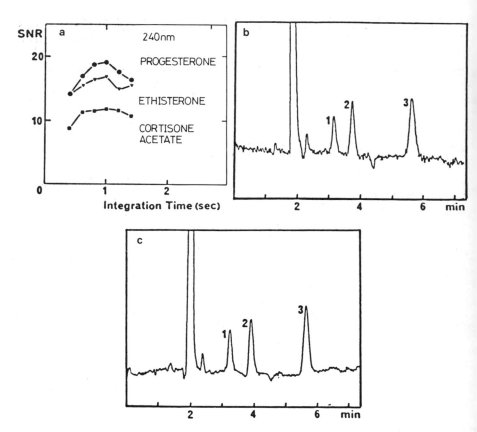

Figure 9 *Improvement in SNR in the time domain for the steroids, (1) cortisone acetate, (2) ethisterone, and (3) progesterone, by increase in the integration time. The separation was on a stationary phase of ODS–Hypersil (Shandon Scientific, Runcorn, UK) with a mobile phase of acetonitrile – water (70:30 v/v). (a) relationship between SNR and integration time; (b) chromatogram at 240 nm for an integration time of 0.4 sec; (c) chromatogram at the optimum integration time of 1 sec.*

(Adapted with permission from A.F. Fell, B.J. Clark, and H.P. Scott, *J. Chromatogr.*, 1984, **316**, 423.)

2.5 Techniques for Testing Chromatographic Peak Homogeneity and Purity

The extra dimension of spectral information from the LDA detector has, through computer aided methods, introduced the capability to identify an analyte by careful use of spectral libraries. In addition, it has permitted the consideration of the resolution efficiency of the HPLC method, through examination of the homogeniety and purity of the chromatographic peak (homogeneous peaks are not necessarily pure).

This latter aspect is of crucial interest in many branches of analytical research. In order to validate chromatographic peak purity and homogeneity it is possible to apply one or more of the many computer aided algorithms which exploit the spectral information acquired by LDA detectors. Numerous methods are available, but it is possible to separate these into two approaches to interrogate the chromatographic peak through use of the spectral information.

2.5.1 Simple mathematical methods. In the first group of methods simple mathematical approaches have been used to give a fast indication of the homogeneity of the chromatographic peak. In contrast the second group of methods as discussed in the next section rely on the more involved, mainly matrix mathematical approaches which can be considered as the chemometric methods. In both cases the methods owe their origins to spectroscopy and statistical methods.

The simplest case of interrogation of the chromatographic peak is to carry out an examination of the spectra at various points on the elution profile. These spectra are generally normalized to allow comparison on overlaying and recorded on the upslope, apex, and downslope which can be quickly executed.

One of the most popular methods which has been the basis of a number of adaptations for peak interrogation is absorbance ratioing. This technique is derived from a classical method developed in analytical spectroscopy and was initially applied in UV–Visible and MS.[16] It is based on plotting the ratio of absorbances at two wavelengths over the elution profile. For a single component with a well-defined spectrum, in the mobile phase, the molar absorptivity $(\varepsilon_{\lambda_1})$ at wavelength λ_1 is directly proportional to the absorptivity ε_{λ_2} at wavelength λ_2.

$$\varepsilon_{\lambda_1} = K_{1,2} \cdot \varepsilon_{\lambda_2} \tag{1}$$

The constant $K_{1,2}$, is characteristic of the pure compound at the wavelengths used. If Beer–Lambert law holds then this constant is independent of concentration and the absorbance value A_1 at λ_1 and A_2 at λ_2 are related through the same proportionality constant

$$A_1/A_2 = K_{1,2} \text{ at } \lambda_1, \lambda_2 \tag{2}$$

Therefore the ratio of absorbances between two wavelengths should be constant and the presence of a partially co-eluting impurity peak, with a different $K_{1,2}$ will cause a change in the resultant absorbance ratio (Figure 10).

[16] V.G. Kostenko, *J. Chromatogr.*, 1986, **355**, 296.

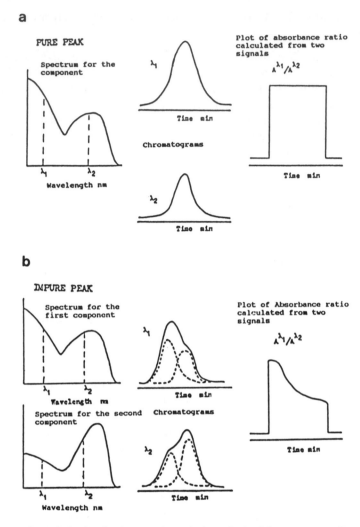

Figure 10 *Schematic for the absorbance ratio method: (a) basis of the square wave form for a pure component; (b) a typical formation arising from an impure peak of different spectral characteristics.*

In the chromatogram the absorbance ratio method (AR) is applied by calculating the ratio at sequential points through the elution profile. With a single pure component a square wave function is given which is equivalent to $K_{1,2}$ across the peak, as in Figure 10.[17] In this example for the trace impurities noscapine and papaverine, which are regularly found in street heroin, the effects on the graphical presentation through the presence of different concentrations is illustrated in Figure 11.

The AR method can provide a swift check for the presence of a co-eluting

[17] A.F. Fell, H.P. Scott, R. Gill, and A.C. Moffat, *J. Chromatogr.*, 1983, **282**, 123.

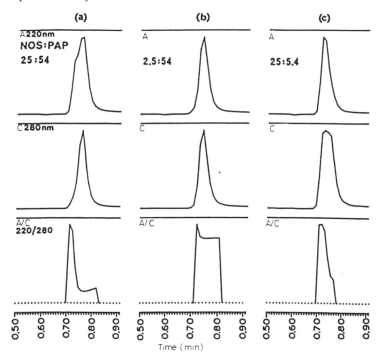

Figure 11 *Overlapping chromatograms at 220 and 280 nm and the corresponding absorbance ratio plots for the alkaloids noscapine and papaverine, which can be found as impurities in street heroin at varying concentration ratios (a) 25:54 μg ml⁻¹; (b) 2.5:54 μg ml⁻¹; (c) 25:5.4 μg ml⁻¹.*

impurity but in practice as the method relies on the dissimilarity of the analytes there can be many drawbacks. These include problems of interpretation due to lack of sensitivity of the method, which is based on the difference in the absorbance ratio between the components. Even in the most favourable case impurity levels much below 10% are difficult to establish. In addition for sensitivity the choice of wavelengths is very important;[18] these can be obtained from spectral information on known materials but are obviously not available for unknown impurities. The AR method is therefore essentially a qualitative tool and a number of modifications have been made which improve its reliability and in addition give a quantitative measurement. One typical example of the former is the Cheng modification.[19]

Another example of an improvement in the AR method is 'Multiple Absorbance Ratio Correlation' (MARC) which can give an indication of purity, detect the presence down to 1%, give a quantitative measure[20] of the impurity, and can be used with more than two coeluting components (which is another

[18] J.G.D. Marr, B.J. Clark, and A.F. Fell, *Anal. Proc.*, 1988, **25**, 150.
[19] H. Cheng and R.R. Gadde, *J. Chromatogr. Sci.*, 1985, **23**, 227.
[20] J.G.D. Marr, B.J. Clark, and A.F. Fell, *J. Pharm. Pharmacol.*, 1988, **110**, 85.

drawback of AR). The method monitors the correlation between the reference spectrum and any other spectrum to give an indication of chromatographic peak purity. It is based on calculating a correlation factor (correlation coefficient, r should approach 1) between the reference data points and the test data points, using an overdetermined system, *i.e.* for 3 components 4 wavelengths are used.

As indicated above simple mathematical methods such as absorbance ratioing are gaining in popularity, with particular emphasis at present on adaptations to present a single figure value to indicate purity. This is considered important, as single figures can be introduced directly into reports, can lend themselves to further numerical evaluation and offer the possibility of being incorporated into automated systems. Single figure methods introduced as commercial products, such as 'Absorbance Index' (Perkin Elmer) and 'Peak Purity Parameter' (Varian) have given considerable improvement upon the AR method. In the index method the absorbance ratio is calculated over the leading and trailing edges of the peak and the values are computed. Regression analysis is then carried out to check the linearity, from which purity is established. With the peak purity par-ameter the single value obtained is reported by the instrument manufacturer to be the absorbance weighted mean wavelength of a spectrum.[21] In both cases indicated above, as in absorbance ratio, the choice of wavelengths or wavelength range examined is important.

A further evolution of the AR method is 'Spectral Suppression' which is a spectral difference method.[22] It is suitable for two components and includes a quantitative measure. The method consists of finding the absorbance ratio at two well-chosen wavelengths for the compounds of interest.

By rearranging equation (1) a null relationship is given:

$$\Delta\varepsilon_{1,2} = \varepsilon_{\lambda_1} - K_{1,2} \cdot \varepsilon_{\lambda_2} = 0 \tag{3}$$

At any concentration the absorbance contribution of a pure compound can be suppressed by computing the difference absorbance function, $A_{1,2}$:

$$\Delta A_{1,2} = A_{\lambda1} - K_{1,2} \cdot A_{\lambda2} = 0 \tag{4}$$

Any coeluting compound with different spectral properties will yield a positive or negative response, whose amplitude is related to concentration.[22] In the biomedical example, in Figure 12, from the chromatogram of some aromatic amino acids and their major metabolites the tyrosine and dopamine peaks appear to the eye to exactly coelute.[23] Although the spectra of these compounds are very similar, there is however sufficient difference to allow the successful application of the spectral suppression method. In order to carry this

[21] A.F. Poile and R.D. Conlon, *J. Chromatogr.*, 1981, **204**, 149.
[22] G.T. Carter, R.E. Schiesswohl, H. Burke, and R. Yang, *J. Pharm. Sci.*, 1982, **71**, 317.
[23] A.F. Fell, B.J. Clark, and H.P. Scott, *J. Chromatogr.*, 1984, **297**, 203.

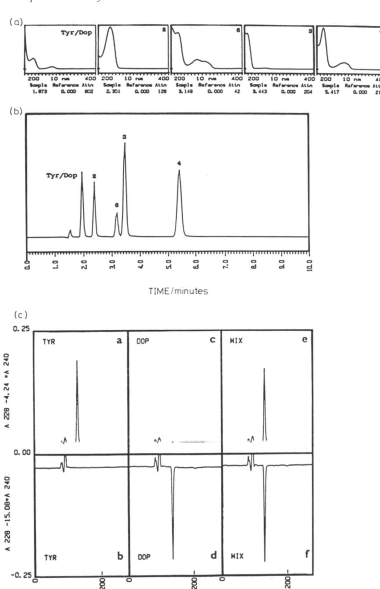

Figure 12 *Use of spectral suppression method in the assay of aromatic amino acids and their metabolites which were separated with an HPLC column packing of ODS–Hypersil and a mobile phase of methanol–phosphate buffer* (pH 4.0) (13:87 v/v): *(a) normalized spectra taken at the peak maxima; (b) chromatogram at 210 nm illustrating the extent of overlap in the peaks for Tyr and Dop; (c) spectral suppression chromatograms for the unresolved Tyr and Dop peak. Spectral suppression at 228 nm and 240 nm. (For suppression equations, see text)*

(Reprinted with permission from A.F. Fell, B.J. Clark, and H.P. Scott, *J. Chromatogr.*, 1984, **297**, 203.)

out the equations to give the suppression of tyrosine (Tyr) and dopamine (Dop) were found to be:

$$\Delta A_{Dop} = A_{228} - 15.08\ A_{240} \tag{5}$$

$$\Delta A_{Tyr} = A_{228} - 4.24\ A_{240} \tag{6}$$

Equation 5 gives a negative peak for dopamine on the suppression of tyrosine and in contrast with equation 6, a positive tyrosine peak on suppression of dopamine. A number of examples of the usefulness of this method have been reported[24],[25] and recently two instrument companies have incorporated algorithms resident within the hardware to allow operation of the method. Spectral suppression, however is only suitable for binary coeluting components and for multiple peak overlap the method of 'Multiple Spectral Suppression' has been developed.[26] In this the spectral suppression method has been extended by using the ratio between the molar extinction coefficients:

$$\Delta A\ (t) = A_1\ (t) - \frac{\varepsilon_1}{\varepsilon_2} A_2\ (t) \tag{7}$$

Where $\Delta A(t)$ is the difference function at a particular time (t) and the molar extinction coefficients $(\varepsilon_1, \varepsilon_2)$ are substituted for A_1 and A_2 in K.

For multiple components the equivalent matrix formula is set up using equation 7 to give:

$$D'\ (t) = \det m = \begin{bmatrix} \varepsilon_2\ A_2(t) \\ \varepsilon_1\ A_1(t) \end{bmatrix}$$

$$\Delta A = \frac{D'(t)}{\varepsilon_2}$$

Where N known components are to be suppressed, D' is a square matrix consisting of a $N + I$ by N submatrix composed of molar extinction coefficients at $N + I$ wavelengths for N known components. In addition there is an $N + I$ by 1 submatrix of total absorbances at $N + I$ wavelengths (where I must be $\geqslant 1$ when N components are suppressed. The determinant $[D'\ (t)\ ]$ (when $I \geqslant 1$) can be calculated from equation 8:

$$D'\ (t) = \sqrt{\det\ (M^T \cdot M)} \tag{8}$$

Where M is a rectangular matrix containing data in a similar format to the above and M^T is the transpose of matrix M. $\Delta A(t)$ can be derived from $D'\ (t)$ using a factor based on the molar extinction coefficients at the relevant wavelengths. This method has been applied to plant extracts from *Catharanthus*

[24] E. Owino, B.J. Clark, and A.F. Fell, *J. Pharm. Pharmacol.*, 1988, **110**, 116.
[25] E. Owino, A.F. Fell, and B.J. Clark, *J. Chromatogr.*, 1990, submitted for publication.
[26] J.G.D. Marr, P. Horvath, B.J. Clark, and A.F. Fell, *Anal. Proc.*, 1986, **23**, 254.

Roseus which provides a source of the antineoplastic alkaloids vincristine and vinblastine (Figure 13). In the chromatogram of a plant extract vincristine and its related impurities overlap as shown in Figure 13a. However, it is possible to suppress each component or group of component peaks to leave a chromatographic peak for the component of interest.

Other methods which come into the category of simple mathematical methods for peak purity assessment include differentiation in the time domain throughout the elution profile. The second derivative has been shown to be the most helpful[17] of the derivative orders, but other higher order derivative methods have also been used.[27] In the second derivative a sharpened bipolar representation of the overlapping components can be produced to improve chromatographic resolution. As with the other methods in this group the analyst has to be aware of the limitations of the method. These include problems when two solutes exactly coelute, as in the case of the aromatic amino acids above, or are separated by less than $0.2 \, w_{1/2}$, where $w_{1/2}$ represents the average spectral bandwidth at half height of the two solutes.

As has been indicated these methods are for swift indications of peak purity, they are not without drawbacks in all cases and in particular they require prior knowledge of at least one of the components in a binary system and generally all the components in a multicomponent coeluting system. If this information is available then the suppression methods can be successfully applied. If a binary system is suspected in the chromatographic peak and both components are known then absorbance ratio or one of its derivatives should give the indication of purity required. Most of the methods are currently incorporated in commercial systems although not many of the systems contains more than two of the methods.

2.5.2 Matrix mathematical methods. The second set of methods can be considered as requiring little prior information about the chromatographic peak; they are generally quantitative, involve more complex mathematical procedures and are often considered as chemometric methods. They are designed to estimate the maximum probable number of components in a peak cluster and with the more advanced methods can yield chromatograms and spectral information on the components. In naming these methodologies there is some confusion over the overall heading for these techniques. In this review the term 'multicomponent analysis' will be used to describe these methods which consist of one or more of the following: spectral deconvolution by least squares or Fourier transform; Curve Resolution (CR); curve fit; Multiple Regression Analysis (MRA); Principal Component Analysis (PCA); Partial Least Squares (PLS) and Iterative Target Transformation Factor Analysis (ITTFA).

One form of spectral deconvolution (curve fit) employs all the wavelength data in a matrix based least squares method[28] where unfortunately all the com-

[27] A.A. Fasamade, A.F. Fell, and H.P. Scott, *Anal. Chim. Acta*, 1986, 187, 233.
[28] M.J. Milano, S. Lam, M. Slavonis, D.B. Pautler, J.W. Pav, and E. Grushka, *J. Chromatogr.*, 1978, 149, 599.

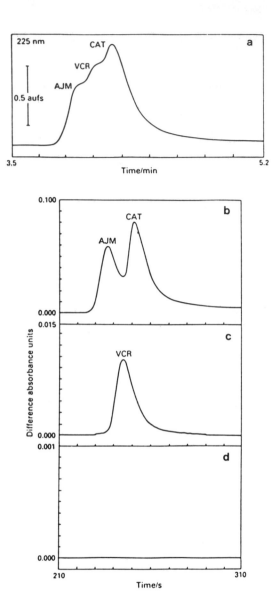

Figure 13   *Multiple component spectral suppression for deconvolution of the overlapping peaks of the antineoplastic alkaloid vincristine (VCR) (84.15 µg ml⁻¹) and its related impurities, ajmalacine (AJM) 89.48 µg ml⁻¹) and catheranthine sulphate (CAT) (85.40 µg ml⁻¹). Partially resolved with a column of SAS–Hypersil and a mobile phase of acetonitrile–ammonium carbonate (46.5:53.5 v/v): (a) overlapping chromatogram at 225 nm; (b) suppression of vincristine to leave ajmalacine and catharanthine; (c) ajmalacine and catharanthine selectively suppressed to leave vincristine; (d) all three components suppressed.*

(Adapted with permission from J.G.D. Marr, P. Horvath, B.J. Clark, and A.F. Fell, *Anal. Proc.*, 1986, **23**, 254.)

ponent spectra must be known. The component spectra concentration coefficients are computed together with the wavelength information from the mixture spectra and from this a composite spectrum which best fits the spectral data is generated. The method is repeated at sequential time points in the elution profile to give a series of deconvolution chromatograms as in Figure 14. In this example for the alkaloid impurities noscapine and papaverine which coelute under the chromatographic conditions used, the method yielded the presence of an impurity previously unidentified.[17]

In carrying out the methods of MRA and PLS it is also most efficient if the spectral information is available for the MRA and PLS and the success of the methods depends on the similarity between the standard and actual spectra in

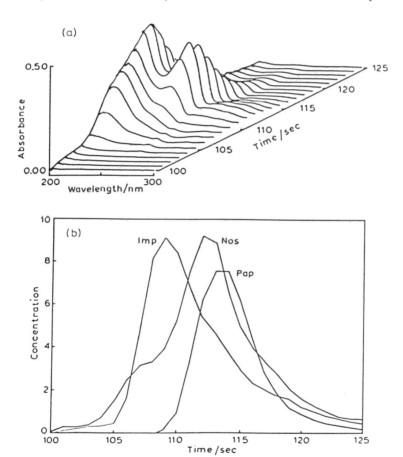

Figure 14 *Least squares deconvolution of the early eluting overlapping impurities, noscapine and papaverine in a heroin sample: (a) pseudoisometric projection of the overlapping noscaprine and papaverine spectrochromatographic data; (b) spectral deconvolution of the above data matrix revealed a third leading edge impurity together with the deconvolution of noscapine and papaverine*

(Adapted with permission from A.F. Fell, H.P. Scott, R. Gill, and A.C. Moffat, *J. Chromatogr.*, 1983, **282**, 123.)

the coeluting peak. Both methods can be performed on overdetermined systems where the number of observation wavelengths at which data are recorded exceeds the number of components in the mixture (it is possible to perform MRA on the same number of datum points). PLS uses all the available information and operates on a curve fitting basis where the differences between known and unknown components in the mixture are minimized by taking the squares of the residuals in the calculation.[29] In carrying out both these methods it is important to place restrictions on the operating data to exclude negative values, linear range, and noise. PLS is often used in conjunction with Factor Analysis (FA) where FA reduces the large data matrix to a more manageable size. FA indicates the number of factors influencing the data set and the nature of physically significant parameters. This is carried out by reducing the data matrix to a minimum which produces a set of orthogonal component vectors that envelop the significant information in the data set.[30] In addition factor analysis has also been reported as an individual technique for interrogating chromatographic peaks. Although MRA, PLS, and FA have been shown to be powerful techniques for peak homogeneity testing, up to the present commercial software packages are not available for use with LDA detectors in HPLC.

Two commercial chemometrics packages are however available, for interrogation of chromatographic data, both of which were developed by Kowalski and co-workers.[31,32] The first of these was developed for a Hewlett–Packard HP85 microcomputer and involves a principal components analysis method (PCA) (MCR−2, Infometrix Inc., Washington, USA). Although the packages are designed to run on multiwavelength data, because of the memory constraints of the microcomputer it only operates on data files from 8 simultaneous chromatograms. The PCA software is essentially a statistically based form of FA and involves a number of assumptions and limitations about non-negative responses, linearity, noise, and minimal difference between the components. PCA involves the extraction of the eigenvalues, usually in association with their corresponding eigenvectors, from a similarity matrix which is calculated from the original data set. The similarity matrix can result from different forms of the data which can be manipulated prior to analysis. In terms of the (A, λ, t) data this can be represented by time delay (rows, spectra) and wavelengths (columns, elution profiles). From this a covariance matrix can be calculated. An eigenvalue is a measure of the relative importance of its corresponding vector. The principal components obtained from the factor analysis are abstract combinations of the chromatograms and spectra from the individual components. By introducing further data into the method which can be in the form of limitations on the data or further spectral information, this then allows transformation (rotation) of the abstract eigenvectors from which meaningful spectral and chromatographic data is extracted.[32]

PCA as set out in the commercial package is generally considered to be limited

[29] M. Otto and W. Wegsheider, *Anal. Chem.* 1985, **57**, 63.
[30] P.G. Gillette, J.B. Lando, and J.L. Koenig, *Anal. Chem.*, 1983, **55**, 630.
[31] M.A. Sharaf and B.R. Kowalski, *Anal. Chem.*, 1981, **53**, 518.
[32] E. Sanchez, L.S. Ramos, and B.R. Kowalski, *J. Chromatogr.*, 1987, **385**, 151.

to < 4 components and the efficiency depends on the setting up of the method to extract the most useful data, the spectral range examined, the similarity of the spectra, and the chromatographic resolution of the coeluting components.

Although the spectral information for the components does not need to be known it does help in the accuracy of the method. In a simple case of a three component overlap (Figure 15) a time window is initially chosen and the data in that window are organized within the computer programme to establish a data matrix of S rows of spectra and W columns of wavelength information (W,

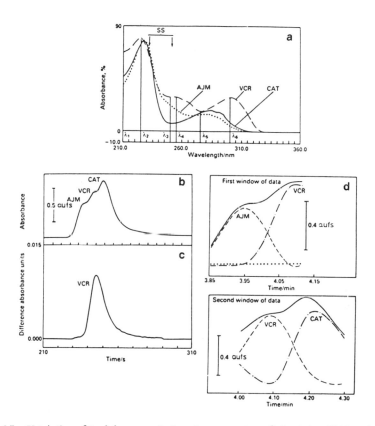

Figure 15 *Validation of peak homogeneity in a ternary system of vincristine (VCR), ajmalacine (AJM), and catharanthine (CAT) by multiple spectral suppression and principal components analysis for chromatographic conditions, see Figure 13); (a) individual spectra of VCR, AJM, and CAT; (b) ternary overlapping chromatogram at 225 nm, absorbance scale 0.5 [absorbance units full scale (AUFS)]; (c) difference chromatogram for VCR by multiple spectral suppression method, with AJM and CAT suppressed; (d) use of a commercial principal components curve resolution package to resolve the two initial overlapping peaks, AJM and VCR and then the latter two peaks VCR and CAT, within different time windows. (For details, see text)*

(Adapted with permission from G.G.R. Seaton, J.G.D. Marr, B.J. Clark, and A.F. Fell, *Anal. Proc.*, 1986, **23**, 424.)

4 or more, *i.e.* overdetermined). After normalizing the spectra to unit area, an initial check for peak integrity is made by plotting all spectra in W dimensions. If the points are concurrent within instrumental noise limits then one component is present. If the programme fails at this point then it continues to derive the first and second principal eigenvectors and eigenvalues $X^T.X$, the covariance matrix of the mean centred normalized spectra, by solving equation 9:

$$(X^T.X) \, V_n = \lambda_n \, V_n \tag{9}$$

where V_1 is the first eigenvector corresponding to λ_1 the largest eigenvalue. If V_1 and V_2 account for $> 99\%$ of the variance, then two components are present within the data set as in Figure 15. Resolution of the components starts by projecting all spectra to a 2-dimensional eigenvector space, so that they lie on a straight line with the pure spectrum of component 1 at one end and the second component at the other. Kowalski[31] showed that the ratio of the 2 components at any time point (i) is equal to the ratio of the so called Euclidean distances of the pure spectra from (i) and from this equation 9, the individual component contributions to the total absorbance, can be obtained.

The other commercial package, also from Kowalski and co-workers[32] features curve resolution (QuickRes, Infometrix Inc.), and runs on a Hewlett–Packard microcomputer (Model 3000). CR has the advantage over the PCA programme that it operates on full spectrochromatographic data and that little information about the components is required. However, as for PCA, it is helpful to obtain the optimum operating conditions, if some information is available. It is essentially a constrained factor analysis method which again requires the choice of a suitable time-window. The method is used to transform abstract spectra or chromatograms from the factor analysis portion of the method into recognizable forms. The developed algorithms use methods which improve upon the traditional tangent-skim or drop-line methods.[31] Generally, however, the method is only suitable for up to three components although this is not considered to be a major problem as the presence of more than three components should not regularly occur. When there is the possibility of more than 3 components then Iterative Target Transformation Factor Analysis (ITTFA) has been shown to be particularly useful and especially when prior information is not available.[34] In ITTFA a target test spectrum is projected into hyperspace and compared with the actual spectral data. The initial process is to check whether the test spectrum is present and then it is possible to carry this one stage further and extract chromatographic information. This is carried out by rotating the abstract profiles to give targets and retesting of the targets until no further iteration is required. If quantitative information is required then the identified components from the test spectrum can be intro-

[33] G.G.R. Seaton, J.G.D. Marr, B.J. Clark, and A.F. Fell, *Anal. Proc.*, 1986, **23**, 424.
[34] B.G.M. Vandeginste, F. Leyten, M. Gerritsen, J.W. Noor, and G. Katemen, *J. Chemometrics*, 1987, **1**, 57.

duced as pure spectra. From this information it is possible to calculate elution profiles by using least squares best fit procedures. As with all the other methods, within certain limitations ITTFA can give accurate quantitation of unresolved chromatographic peaks. The limitations again include the sensitivity to chromatographic resolution and the degree of spectral similarity. It is, however, one of the most powerful techniques for deconvoluting spectro-chromatographic data which can require no prior knowledge, and gives estimates of spectral and elution profiles together with quantitative information. Numerous multicomponent analysis methods exist for LDA detection in HPLC, of which the above are just a small number. In the order presented here the degree of complexity is increased through to ITTFA which requires a powerful 32 bit or mainframe or mini computer for operation. Following alongside the methods is the degree of need of prior information and in that case methods such as ITTFA which requires no information are very powerful.

At present one of the major problems for the analyst is the lack of suitable commercial software to address the problem of chromatographic peak homo-geneity and purity. It is to be hoped that suitable algorithms will follow the rapid advance in computer hardware thus allowing the incorporation of more of the chemometric methods, and especially those where little prior information is required.

2.6 Expert Systems

The value of expert systems in HPLC are only just being realized. At present, reported expert systems have concentrated on one aspect of the chromato-graphic variables, *i.e.* column selection,[35] mobile phase choice and optimiz-ation,[36] and chromatographic peak purity assessment.[37] In the latter two aspects multichannel detection can offer considerable advantages in an auto-mated expert system set up. In the case of optimization of mobile phase the complex methods now in use, simplex, resolution mapping, and window diagrams[9] can be incorporated with peak tracking methods and peak homo-geneity methods in one 'intelligent' software package.

3 CONCLUSIONS

It has been shown that the multichannel detector has much to offer the analyst in terms of improvements in selectivity through spectral recognition and peak purity assessment. A new area has opened up in the assistance provided in the setting up of optimization strategies which are becoming more and more important as the range of separation mechanisms, column packing materials, and complexity of methods expands. The concentration has up to present been towards the absorptiometric multichannel detector, however, we must

[35] G. Musch, M. De Smet, and D.L. Massart, *J. Chromatogr.*, 1985, **358**, 97.
[36] A.F. Fell, T.P. Bridge, and M.H. Williams, *J. Pharm. Biomed. Anal.*, 1988, **6**, 555.
[37] T.P. Bridge, M.H. Williams, and A.F. Fell, *Anal. Chim. Acta*, 1989, **223**, 175.

not dismiss the multichannel possibilities in luminescence which, although at present are limited to one complete commercial unit, can present another dimension in sensitivity and selectivity.[38] Multichannel detection in HPLC is surely set for a number of years of further expansion especially as the computing powers available are continually advancing together with suitable software developments. In tandem much more interest is likely in the coupling of the other spectroscopic methods, with interfacing formats being improved, and more commercial considerations developing.

[38] B.J. Clark, A.F. Fell, and D.G. Jones, *J. Pharm. Biomed. Anal.*, 1988, **6**, 843.

CHAPTER 7

A Practical Approach to Quantitative Methods of Spectroscopic Analysis

J. COATES

1 HISTORICAL PERSPECTIVES AND GENERAL BACKGROUND

After studying chemistry at college, or even in high school, some of our earliest recollections of quantitative analysis by spectrometric methods might be the measurement of sodium or potassium content in an aqueous solution by flame photometry, or the determination of phosphates via phosphomolybdate blue by visible photometric measurement. In both cases a series of standards were produced, often made by serial dilutions, and a reading was obtained for each standard at a given wavelength, usually defined by an optical filter. A calibration graph was constructed, manually of course, by plotting the instrument reading *versus* the concentration of the species being determined. The sample, or 'unknown' was then measured under identical conditions, and the analyte was measured by 'reading off' the concentration that corresponded to the reading on the graph. This was fine because it taught us that light interacts with matter, and if we do the experiment carefully, the degree of interaction is linear with the change in concentration of the analyte. Unfortunately, it told us very little about spectroscopy and the true nature of the interaction of light with matter, also it did not teach us anything about the potential pitfalls of quantitative analysis by spectroscopic methods.

It can also arise, that we become biased in our thinking such that we believe that quantitative measurements are limited to ultraviolet or visible methods of analysis. An appreciation for the general concept of interaction of electromagnetic radiation with matter is seldom gained until later. Consequently, to some people, quantitative analysis is as simple as following a cookbook method, which involves diluting the sample in a suitable solvent, often water, and pouring the solution into a glass cuvette. In such instances people do not consider infrared spectroscopy or nuclear magnetic resonance spectrometry as candidates for quantitative analysis. If we leave the discussion of quantitative analysis to a coverage of only the ultraviolet – visible methods we will fall short on our understanding of the laws of nature that are involved, and we will miss the role that the sample can play in the analysis. In order to cover the practical, as well as theoretical aspects of the subject, we will use the mid-infrared region of the spectrum as a model for our discussion. In many ways, this region of the spectrum can be one of the most difficult to handle; this may result from the nature of the sample or the physics of the experiment. However, if we persevere, the result can be most rewarding. Not only can we potentially carry out analyses which would be impossible by any other technique, but we can also obtain a lot of information about our sample in the process. Although the discussion in this text will be focused on infrared methods and problems associated with infrared sampling, and although the nature of the primary data and the methods of measurement are different, it must be realized that many of the concepts and methods discussed carry over into other forms of spectroscopy.

Probably the most obvious difference between the techniques of UV–visible and infrared spectroscopy is the appearance and the information content of the spectra. A UV–visible spectrum of a compound in the condensed phase typically exhibits one or two broad and diffuse absorption contours. Each contour may be complex but usually the features are severely overlapped and unresolved. This tends to simplify the choice of how to define the parameters required to make a quantitative measurement. The infrared spectrum, on the other hand is very rich with information, typically containing a large number of relatively well resolved peaks. These peaks, or absorption maxima, are specific to vibrations associated to different functional groups or structural features of the molecule. This tends to complicate the selection of the parameters used for quantitative analysis especially as interactions may occur between different functional groups thus changing the molecular environment of the compound being measured. Consequently, not all parts of the spectrum necessarily respond in the same quantitative manner to changes in concentration. Deciding what information to use is part of the method development procedure, and this will be covered later in this text. Once the analytical parameters are defined, in terms of the peak or peaks of interest, the analysis then returns to the same common ground as the UV–visible method. Methods for the numerical extraction of data from the spectrum and the methods of manipulating the data to provide the final result are common to both UV–visible and infrared spectroscopy. It is interesting to note that although manual methods for UV–visible analysis have been passed down for decades, the first commercial implementations of computer-based quantitative analysis were directed towards infrared methods. In fact today, near-infrared and mid-infrared spectroscopy lead the way in the development of sophisticated computer-based software for quantitative analysis.

To return briefly to a historical perspective. If we go back to the beginnings of analytical infrared spectroscopy, we find that some of the earliest industrial applications of the technique were quantitative methods. These were developed in response to the need to perform quality control of synthetic rubbers and petroleum fractions. Most of the early work was performed at large chemical companies. In fact, the need was so strong for the analysis that the initial developments on analytical infrared instrumentation were carried out at large chemical companies, such as Shell, Dow, and American Cyanamid. In all of these cases the instruments were developed into commercial products. The early measurements were rudimentary, and involved the generation of a single beam spectrum. It is not uncommon nor difficult to use a single beam measurement for UV–visible measurements. In those regions of the spectrum where there is no interference from atmospheric absorption (water vapour and carbon dioxide) single beam measurement may also be used for quantitative analysis. However, more generally infrared measurements are made from a 'double beam' spectrum, where the single beam background of the spectrometer and the atmospheric absorptions are ratioed from the sample spectrum. Double beam infrared instruments were available by the mid-1950s. Although some of the earliest applications of these instruments were for quantitative measurements, the instrumentation was far from ideal for the analysis. Consequently the technique fell out of favour and gas chromatography took its place for routine quantitative analysis. Even as late as 1970, infrared was still considered to be a 'qualitative only' technique. There were two main drawbacks to the infrared instrumentation available at that time. The technique was considered to be noise limited, especially when compared to UV–visible instruments, and the data extraction from the spectrum was crude. The only available form of the spectrum was a chart recording, presented in percent transmittance, a function not linearly related to concentration. Typically a baseline was determined visually and constructed with a ruler and pencil. The transmittance information was read from the peak and the baseline, and the values obtained were manually converted to absorbance. A baseline corrected value was then obtained, and from that point on, the traditional method of using a calibration graph was used to determine concentration.

Major changes occurred in the 1970s that completely altered the role of infrared spectroscopy in quantitative analysis. The two most significant of these changes were the introduction of FTIR (Fourier transform infrared) spectrometers, and the use of dedicated computers with instrumentation. The FTIR spectrometer has provided us with a tool to obtain higher quality spectra under conditions of high signal-to-noise, such that infrared spectroscopy is no longer considered to be an energy limited technique. An on-line dedicated computer provides a convenient means for extraction of the necessary spectrum parameters directly from the digitized form of the spectrum. Data conversions and calibrations can also be handled automatically, as part of an overall computer-based scheme for quantitative analysis. The early FTIR instruments and computers were large, and their use and application was directed towards a research environment. During the 1980s there were significant improvements in spectrometer design and there was a general scaling down in size of the instruments and computers. the trend is now towards desktop FTIR instruments with either a desk-top computer workstation, or a built-in computer. Another key improvement has been in the area of sample handling. It can be said now that the greatest errors in quantitative spectrometeric measurements are user related and sample related, and not instrument related. Although sample handling has improved, it is still one of the most critical steps in the quantitative procedure.

This background information has been provided to indicate that although all spectrometric methods involving electromagnetic radiation are inherently quantitative, getting to the final answer is not necessarily straightforward. This text will now go through the basic concepts and methods involved in modern methods of quantitative spectrometric analysis.

2   THE FUNDAMENTAL CONCEPTS

The statement has been made that spectrometric methods are inherently quantitative. Let us investigate what this means and define the basic laws that govern the measurement. The fundamental premise is that as the concentration of a species increases, there is a corresponding increase in the spectral response. The general prerequisite for spectrometric quantitative analysis can be defined as follows:

> 'Information derived from the spectrum of a sample is related in mathematical terms, to changes in the level(s) of an individual component, or several components, within the sample or a series of samples. That is the spectral response of an analyte can be related by a mathematical function to changes in concentration of the analyte.'

The ideal situation is where the measured spectroscopic feature varies linearly with changes in concentration. In the past, this has always been considered to be essential for good quantitative analysis. In reality, true linearity is not always obtained, but this is not important, as long as the measured function is reproducible. Most practical analyses are not absolute measurements, and normally measurements are made on a given instrument within a fixed working environment. Under these controlled circumstances, reproducibility and consistency of the measurement are the most important factors.

The fundamental relationship that is used to describe the relationship of spectral response to concentration is the Beer–Lambert–Bouguer Law—sometimes shortened to Beer's Law or the Beer–Lambert Law. The basis of these laws is that the absorbance (in the past also known as optical density) of a measured absorption band is a function of the thickness (or optical pathlength) of the sample and the concentration of the species being measured. Originally the laws were stated in two parts:

Lambert's Law (originally stated by Bouguer)

$$a = k_1 l$$

where a is absorbance, k_1 is a proportionality constant, and l is the thickness.

Beer's Law

$$a = k_2 c,$$

where a is absorbance, k_2 is a proportionality constant, and c is the concentration of the species being measured.

The combined law is usually quoted, and is presented as:

$$a = k.l.c,$$

where a, l and c are as defined above. The combined proportionality constant k is a property of the species being measured, and is known as the Absorptivity (or in the past, the absorption coefficient). If the concentration is quoted in moles per litre, and the pathlength is fixed to 1 metre, then the quantity is a fundamental term, known as the Molar Absorptivity. In theory, if this is previously determined for a given species, and the pathlength is known, the concentration can be directly calculated for an unknown from the measured absorbance. In practice this is almost never the case. For infrared measurements, this parameter is very difficult to obtain. It is strongly influenced by the nature of the sample, the method of sample preparation, instrument conditions, and various sources of error that can lead to non-linear behaviour.

The Beer–Lambert–Bouguer Law as stated should always lead to a linear relationship. It does, however, make certain assumptions in the method used for the measurement. It assumes that a parallel light beam (a collimated beam) is used, the outside faces of the sample or sample cell are optically flat, the sample is homogeneous, and that the source of radiation is monochromatic. The first two of these assumptions imply that all the rays of light passing through the sample have identical pathlengths. In practice very few of these assumptions hold true. Most modern spectrometers operate with a focused beam, and all have a finite optical bandpass, which means that there can never be pure monochromatic radiation. In the strictest sense this would imply that it is impossible to perform quantitative analysis that conforms to Beer's Law. For practical situations, however, as stated earlier, no analysis is performed as an absolute measurement, and all evaluations are made on a relative basis against calibrations made on the same instrument for a given set of instrumental parameters. At this point the question could be raised regarding the transferability of quantitative methods between instruments. The concept is feasible on modern instruments, as long as the method is robust, the sampling is well under control, and that adequate cross-correlation and re-scaling of spectra can be performed.

One small point before leaving the discussion on Beer's Law. Sometimes people will refer to a system exhibiting a non-linear Beer's Law response. In the strictest sense, by definition, Beer's Law defines a linear effect. The use of the term non-linear Beer's Law is a misnomer, and it really should be translated as being a non-ideal system, or in other words, the system does not obey Beer's Law. This does not imply, however, that the system cannot be used for quantitative analysis. If the non-linearity is reproducible, and a calibration can be established, then quantitative analysis is possible.

As mentioned earlier, infrared instruments produce a spectrum where the spectral response is presented in units of percent Transmittance, sometimes abbreviated to %*T*. This is the normal form of the spectral data that is generated via the detector of the instrument. It is the ratio of the intensity of the light transmitted through the sample (*i.e.* light that reaches the detector) compared to the intensity of the radiation from the source incident on the sample (*i.e.* light that would reach the detector in the absence of the sample), expressed as a percentage, *i.e.*

$$\text{percent Transmittance} = I/I_0 \times 100,$$

where *I* is the intensity of light transmitted and I_0 is the intensity of the light incident on the sample.

The spectrum is normally presented as a double beam format, where the ratio of I/I_0 is generated by the spectrometer. Modern FTIR instruments are, by design, single beam instruments. As previously mentioned this form of the spectrum is less desirable, but the problem is resolved by a ratio with a previously recorded background spectrum. This is normally handled automatically as a standard procedure within the instrument's software. The relationship between spectroscopic data and component concentration, as defined by Beer's Law, is that absorbance, *a*, not transmittance, is proportional to concentration. Transmittance is related to absorbance, as a logarithmic function—

$$a = \log_{10}\frac{1}{T} \quad \text{or} \quad \log_{10}\frac{100}{\%T}$$

The differences between the transmittance form of spectral data and the absorbance form are important to appreciate. It is very easy, with modern computer-based spectrometers to perform a host of numerical manipulations on a spectrum, often at the push of a button. One must appreciate, however, that almost all numerical operations should be performed on the absorbance form of the data, not the transmittance form. Any attempts to carry out significant numerical processing on the transmittance form of the data can result in the loss of quantitative integrity across the spectrum, which can result in gross photometric distortions of the data. There are some mathematical operations that are legitimate to carry out on the transmittance form of the data, such as spectral smoothing and signal averaging. In both cases, the object of the exercise is to reduce the contribution of high frequency noise in the final spectrum. Noise varies linearly between 100% and 0% transmittance and therefore attempts to perform a smoothing operation on the absorbance form of the data can lead to unexpected logarithmic distortions of the data.

It is reasonable to ask why, if absorbance is so important, do infrared instruments always present the primary form of the data as percent transmittance? The main reason is historical—UV–visible instruments have always been considered to be quantitative instruments, and therefore the conversion to absorbance has always been built into the electronics of the instrument. Infrared instruments were always considered to be qualitative instruments and therefore the absorbance form was not considered to be necessary. Also, from a qualitative point-of-view, the %*T* format is preferred because it portrays signal/noise ratio realistically, without the sacrifice to weak spectral features close to the baseline. On modern instruments the interchange between the two forms of the data is provided as a standard function.

3 DEFINING A METHOD

Having defined the basics, and indicating certain areas where problems might occur, it is now appropriate to outline the key considerations for establishing a quantitative method. In this section, a general discussion will be provided that covers most aspects of quantitative analysis. Details of more specific issues will be covered under separate sections.

The following is a list of the practical steps involved in the development of a quantitative method from basic principles:

   (i)   Understand the system/sample
   (ii) Determine the best method of sampling
   (iii) Prepare standards
   (iv) Run standards and calibrate
   (v) Prepare a validation sample(s) and evaluate the method
   (vi) Analyse samples

The first step of understanding the sample and the chemical system is probably the most crucial for the ultimate success of the analytical method. Too often the statement is made that a particular sample cannot be analysed because the first attempt to do the analysis failed. A common reason for failure is a lack of understanding of the sample, the sampling matrix, and possible interferences. Traditional UV–visible quantitative analysis might appear to be simple, and generally not prone to such problems. For many analyses, the sample is dissolved in a suitable solvent, and the absorption from the component is measured directly. This often yields what can be considered to be a single component system. Most common solvents used for UV–visible analysis are essentially transparent throughout the spectral region that is used for the analysis. Also, unless a second component in the system is known to be a UV absorber, there is little chance of a major interference from any other components in a sample. A good example could be a liquid analgesic containing a single active ingredient. Typically these are syrups composed of the active ingredient, sugars, and other incipients, plus water. This could pose as a simple analysis for a UV–visible method because there are no significant interferences from the matrix. On the other hand, for infrared analysis, this could be a difficult sample.

Without exception, there are no ideal solvents for infrared spectroscopy. All materials, except certain gases, have an infrared spectrum. At one time materials such as carbon tetrachloride and carbon disulphide were touted as good infrared solvents because they had large regions of transparency over the full range of the infrared spectrum. These materials were limited in application because their solvent powers were not strong, and they dissolved only a limited range of materials. As a rule of thumb, materials that have good solvent powers, such as water, acetone, dimethyl formamide, methanol, *etc.* are generally considered to be poor infrared solvents because they exhibit strong infrared absorption over major areas of the spectrum. As a secondary issue, the use of carbon tetrachloride and carbon disulphide is now considered to be undesirable and their general use is prohibited in several countries. The general point to be made here is that almost any system that is studied by an infrared method of analysis must be considered as a two component system (at minimum). Therefore, strictly speaking, there is never a situation where an analysis can be considered as a single component system. It is essential to consider the role of the solvent and other possible constituents while defining the method.

To return to the example of the simple liquid analgesic. The major ingredients in this type of product are water, sugars, preservatives, and sometimes ethanol. Minor ingredients are the active component, flavourings, and colour. The infrared spectrum of this material will be dominated by the main solvents, water and possibly ethanol, with significant contributions from the sugars. It is probable that some spectral features can be assigned to the active ingredient, and therefore a method could be defined around the measurement of the component. In this case, understanding the sample is essential because the contribution from the matrix is a major factor, and can pose as a major interference. Also, there could be chemical or physical interaction between the active ingredient and the solvents resulting from solvation and/or hydrogen bonding. These can also have a major influence on the analysis, and often account for non-linear effects.

It is reasonable to ask at this point—why not use UV–visible for this particular analysis, rather than experience all these potential problems by an infrared method? If the requirement is to measure only the level of the active ingredient, and if the active ingredient is known to be correct, then a UV–visible method is probably the best procedure. If, however, a more complete assay is required, then the infrared approach may be the preferred method. The components that could be considered to be interferences, could also be considered to be analytes. If this is appreciated ahead of time, then the method can be set up as a multicomponent analysis where all major ingredients and the active ingredient are determined simultaneously. Also, because of the increased selectivity of infrared spectroscopy, there is the potential to provide confirmation that the correct active ingredient has been added. UV–visible spectroscopy would be weak in both of these areas.

Therefore before starting the development of a method it is important to: define what is required from the analysis—in the example provided—whether it is the single active ingredient, or a complete assay, determine the nature of the material to be analysed, establish the spectral contribution of other ingredients, and evaluate the possibility of mutual interaction between ingredients.

Once the correct approach is established, it is important to obtain reference spectra of the component being analysed, and any other significant ingredients. The spectral regions of interest are then defined, and the most likely peaks and baseline points (if pertinent to the final method) to be used for the analysis are identified. At this stage, an integral part of the method development procedure is to decide on the most efficient, and reproducible sampling procedure, and to optimize the instrument measurement parameters. This is carried out in conjunction with the spectral evaluation, to ensure that the correct sampling procedure is used to provide the necessary sensitivity, to ensure that the spectral features are adequately resolved, and to ensure that the best (or optimum, timewise) signal-to-noise performance is achieved.

Once the concept of the method is established, an adequate set of standards is required. The composition and number of standards required is dependent on the complexity of the chemical system to be analysed. In the past, crude attempts at quantitative analysis were made on simple photometers by the use of a single calibration standard. The use of a single standard assumes that there are no measurement errors, that the standard is accurate, and that the calibration for the species involved is linear and has a zero intercept. This is generally not considered to be good practice, although the approach is some-times still used for certain methods of elemental analysis. Careful attention to the preparation of standards is essential because errors can be compounded. For some methods the quality of the standards can mean the difference between success and failure. Specific issues on standards will be covered later in this text. Producing a calibration is not the end of the story because the method must be tested before it can be considered to be successful. The method is tested by the use of validation samples. Validation samples are sam-ples of known composition, prepared in the same manner as the standards, but are not included in the calibration set. These are used to provide statistics for the overall method, and to test the accuracy of the method over the range of the calibration. After testing with an appropriate number of validation samples, the method is ready for the analysis of a sample. At each stage, the sampling and instrumental procedures used for the calibration and validation stages must also be used for the samples.

4 CLASSIFICATION OF CHEMICAL SYSTEMS

In a traditional approach to quantitative analysis, the methods are often divided between single component and multiple component systems. This is not a particularly practical approach because many systems that are consid-ered to be single component systems are better served by one of the methods used for multi-component analysis. Conversely, some systems that by nature are defined as containing more than one ingredient can be adequately handled by procedures used for classical single component analysis. In this text the terms simple and complex systems will be used. Simple systems are chemical systems containing one or more components, that can be accurately analysed (within defined limits of expectation) by simple, two dimensional methods of numerical evaluation. That is, the calibration for the analyte can be repre-sented on a simple two dimensional plot. This does not imply that the plot must be linear, or that the plot has to adhere to confines of the Beer–Lambert–Bouguer Law. Complex systems are ones where the simplistic approach of the two dimensional model does not adequately fulfill the needs of the analysis. Typically this applies to multi-component systems where there is a significant degree of spectral overlap, or where there is a large amount of mutual inter-action (physical and/or chemical) between the components. There are cases where the simple approach works adequately over a short range of concentrations, but becomes inadequate as the spread of concentrations increases. Other situations that are better approached as complex systems are where the absorptions of the analyte (or analytes) are low relative to the background absorption of the matrix or solvent. A similar situation occurs when the signal-to-noise level obtained for a particular analytical procedure is less than ideal.

5 SIMPLE SYSTEMS

Simple systems can range from simple mixtures of solvents, to polymer systems involving additives and copolymers, to simple gas mixtures. By the definition provided, it is necessary to identify an absorption or another characteristic of the analyte that can be measured and calibrated by a reproducible procedure, and that can be represented in a simple graphical (two dimensional) manner. Some typical samples that can be considered as simple systems, are simple solvent mixtures, such as paint thinner, copolymers and polymer blends, additives in polymers (slip agents, antioxidants, and plasticizers), additives in lubricating oils, and *trans*-unsaturated components in natural oils and fats. In all of these cases, the components of interest can be clearly identified from their infrared spectra, even in the presence of other materials. The examples provided illustrate different scenarios: in the first two, all of the components are present at high concentration, the second two involve minor components, and the final example involves the measurement of a specific functionality, rather than a discrete component.

The first stage in the development of the method is to be able to measure adequately the spectral feature, usually an absorption band, associated with the analyte. Absorption band evaluation can involve more than just the measurement of peak height. While for many analyses, the measurement of peak height may suffice, it is not necessarily the best or only choice. The determination of the peak itself may be obvious, but on the other hand, the selection of a baseline is not necessarily as obvious. The main problem arises from the fact that the infrared spectrum of any material is relatively cluttered with absorption bands. Even though a peak may appear to be isolated from its neighbours, it is still influenced by their close proximity, especially in the area of the base of the peak. One of the main problems is that the concentration of another component in the system may be changing by a larger amount than the concentration of the analyte. On first inspection it may appear that the absorptions from the other component are far removed from the analyte band. However, it is a common occurrence that the wings of a distant band can influence another band up to 100 cm^{-1} away, or that a minor absorption associated with an interfering component is located close to the base of the analyte band. In such cases, the selection of a baseline point far removed from the analyte band might be a better choice. The only danger from the use of a remote baseline point is that other sample-related factors, such as the effect of the sampling technique, could independently influence the baseline measurement.

One of the best approaches to baseline selection is to carry out several calibration runs and to evaluate the quality of the calibration. Usually the best selection is the peak and baseline combination that produces either a linear or a near-linear plot with a zero intercept. A large offset from the zero intercept might be indicative of a bias caused by a poor selection of baseline. Another aspect of baseline selection is how it is constructed. Traditional thinking often errs toward the use of a baseline tangential to the base of the peak. In practice this is not necessarily a practical or the best approach. Figure 1, the absorbance spectrum of a solution of tung oil-based alkyd resin in a petroleum solvent (a commercial wood finishing agent) helps to illustrate some of the potential problems. Often, when dealing with an envelope containing several bands, a single point baseline, removed from the envelope, such a point close to 2000 cm^{-1} (Figure 1) can be a good choice when dealing with relatively intense bands, such as the carbonyl band at 1740 cm^{-1} (A), or the methylene (-CH$_2$-) band at 1465 cm^{-1} (B). In the case of a weaker band which is superimposed on a relatively strong background, such as the minor *trans*-unsaturation band at 965 cm^{-1} (C), a choice of a pair of points either side of the peak would be the best selection because they would be more representative of the background than a remote point. It should be noted that in UV–visible measurements the problem is invariably less acute unless the system is a complex mixture. On many occasions, the net absorbance of the analyte band obtained from a solution in a suitable solvent relative to the solvent blank is sufficient.

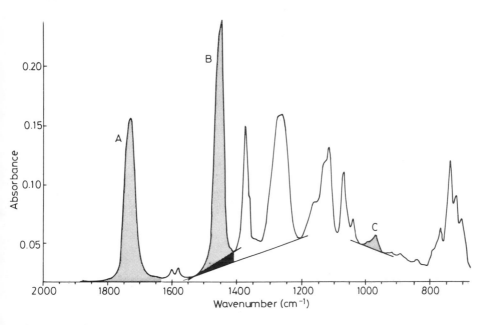

Figure 1   *The selection of peaks and baselines in a spectrum*

The selection of an ideal band for evaluation may be straightforward if the spectrum is relatively simple, as in the case of Figure 1. At this stage, a peak height based measurement can be made as described above. However, an alternative method is the use of integrated absorbance, or peak area, for the evaluation of the peak. This method, in the past, has been proposed as a superior method to peak height evaluation because it can yield improved precision, and it is less sensitive to small variations in spectral resolution. These two issues were more critical on older dispersive instruments where the signal-to-noise performance was relatively poor, and minor changes in resolution could be experienced from variations in the mechanical slit mechanisms. When using modern instruments, where neither situation is an issue, peak height measurements are frequently as reliable as integrated peak areas. It should be noted, that when using integrated areas, the baseline points used for a peak determination are not necessarily the best for area measurement. Large amounts of 'non-peak' information, which may result from the use of an excessive amount of baseline or background information, contribute to the noise content, rather than the signal (peak information) content of the result. One solution to reduce this effect of added noise is to integrate a slice through the peak that includes the heart of the peak but excludes the wings of the peak. With this method of data extraction it is still necessary to define a baseline, as indicated in Figure 2 (where A_1/A_2 define the area limits, and B_1/B_2 the baseline limits).

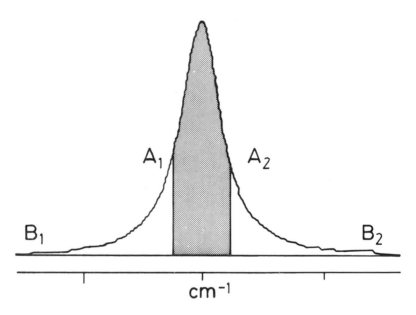

Figure 2   *The evaluation of peak area*

The most suitable analytical band is not always the most significant absorption band in the spectrum. This is why it is essential to understand the sample and its spectrum over the full range of concentrations. For example, problems can result from the absorption band profile changing with sample concentration, or where severe overlap from a neighbouring absorption occurs. Some of the common causes of band profile distortion are mutual interaction effects, such as solvation, and hydrogen bonding effects. Attempts to calibrate on bands that exhibit interaction effects will result in a non-linear response. Similarly, attempts to measure a partially resolved band on the side of a second, major band can also lead to non-linear calibrations, and sometimes unpredictable behaviour if the sample contains more than two components. One method of improving the linearity in an overlap situation is to use the first or second derivative form of the spectrum. The first derivative of an absorption has a positive and a negative lobe as shown in Figure 3 (positive lobe shown shaded, L_1). Selection of the lobe furthest from the interfering band can often lead to an improved calibration. A measurement may be made by evaluation of the span between the top of the lobe, L_1 and the baseline, B_1. Further improvements may also be gained by the use of the second derivative function. Because of the increased level of noise content of derivative functions, the use of neither function is recommended if the band overlap is not excessive.

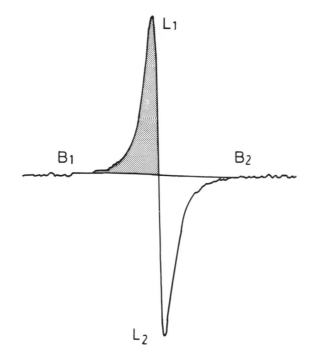

Figure 3   *The first derivative of an absorption band*

The preparation of an adequate number of standards covering the anticipated concentration range is the most critical part of the analysis. There may be a temptation to make a single standard and then to perform serial dilutions to produce a series of standards. This is not good analytical practice because it relies on a very high accuracy for the production of the first standard. Any error in the preparation of the first standard will be carried over into all the other standards (plus any dilution errors). There must be a sufficient number of standards to be able to provide statistical significance to the calibration, and adequately define any non-linearities. The number of standards required is dependent on the nature of the analysis and the number of components to be analysed. Figure 4 illustrates a pair of typical calibration curves—one obeying Beer's Law (A), and the other not obeying Beer's Law (B), *i.e.* exhibiting deviations from Beer's Law.

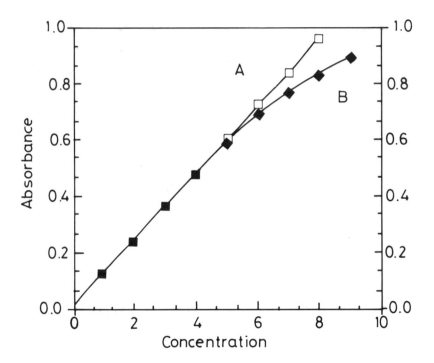

Figure 4   *Beer's Law, (A) compliance, (B) non-compliance*

Until this point, there has been an assumption that the pathlength of the standards is known, or at least that all of the standards are measured at constant pathlength. This condition is practical for liquid samples and solutions. However, solids in the form of powders and extended surfaces, such as polymer films and sheets, can create practical difficulties when it comes to maintaining a constant sample thickness, or a known sample thickness. Furthermore, sample preparation methods for solid samples have practical limitations in their ability to provide reproducible sampling. The most straightforward samples to handle are self-supported polymer films because the thickness can be measured with a micrometer. For cast polymer films, this is not practical, and alternative computational methods have to be adopted. Two practical approaches are a band ratio method, and an internal standard method.

The band ratio approach requires a reference band to be selected, which is related to a major component in the system, but is unrelated to the analyte. Good examples are the determination of plasticizers in plastics films, or the determination of combined vinyl acetate in EVA (ethylene-vinyl acetate) copolymers. In both of these cases, the determination of the ratio of the carbonyl band of the plasticizer (or the vinyl acetate) against a base polymer band, such as the methylene (-CH_2-) deformation, would provide a suitable method for measurement. The premise held for the band ratio method is that the intensity of all bands in the spectrum vary linearly with change in pathlength, and vary by an equivalent amount. By calculating the ratio of the absorbance of the analyte band to the absorbance of the reference band, the contribution of the pathlength, in the Beer's Law calculation, is effectively cancelled out. A calibration for the standards is constructed from the absorbance ratios *versus* the concentration ratios for each standard.

In a multiple component system, such as a terpolymer, it is possible to apply the reference band approach to the simultaneous determination of all three monomer species. The method can involve a reference band selected from one of the components being analysed. The only criterion that must be held for the method to work, is that all the components being measured must add up to a known fixed amount. For the terpolymer example, with three ingredients being determined (and no other component being present) all three components would add up to unity (100%).

The internal standard method is a variant on the band ratio procedure. In this case, another material, which is not part of the system being analysed, is added, at a known concentration to the standards and the sample. A knowledge of the concentration of the reference band permits the back-calculation of an equivalent to the pathlength which can be applied to the standards and sample alike. This internal standard approach is less suitable for polymer films, but works well when powdered samples are being analysed. For all methods that use a band as a thickness reference, it is essential that the reference band selected is well defined, preferably isolated from other absorptions, and free from interference.

Sometimes, it is difficult to build a set of standards from the individual ingredients. A good example would be the measurement of a particular ingredient in a natural product, such as the determination of caffeine in coffee, or the measurement of glucose or fructose in a natural syrup. A procedure known as the method of standard additions, used extensively in atomic absorption measurements where matrix effects are important, can be used. In this method, the material being analysed is added in known amounts to the sample. After each addition the spectrum is recorded and the absorption band of the analyte is measured. A unique calibration for the sample is generated by plotting the quantity of standard material added *versus* the intensity of the analyte. The concentration of the analyte already in the sample is determined by back extrapolating the calibration until the line intersects the concentration axis (X), as illustrated in Figure 5.

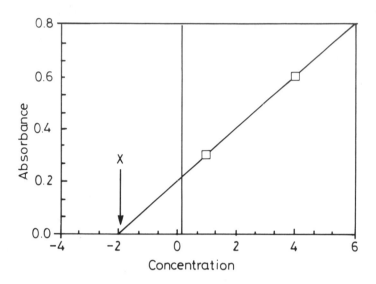

Figure 5 *The method of standard additions*

6 COMPLEX SYSTEMS

As defined earlier, a complex system is a chemical system containing two or more components, where the components are not accurately determined by the simple, two dimensional approach. Typical systems involve spectra where there are several components but the bands of the components are not all well defined in the resultant mixture spectrum, or where there is significant overlap or interference between the peaks that would be selected normally for the individual components. In this second case there may only be two or three components but they may be chemically similar and not adequately differentiated in the spectrum of the mixture. With a traditional approach to quantitative analysis, the discussion is either centred around single component or multicomponent systems. Based on the traditional approach, the measurement of low levels of a sugar in water might be considered as a single component system—an analyte dissolved in a solvent. In practice, however, because the water spectrum is very intense, and because there is a relatively large solvation/ hydrogen bonding interaction between the carbohydrate and water, it can be better to consider the solution as a complex system.

In a traditional approach to multi-component analysis, the contributions of every component in the system are accounted for in the evaluation of each component. A simplistic method for providing this 'compensation' is by expressing each peak absorbance as a summation of all the contributing absorbances from each component, and then to solve for each component by a series of simultaneous equations. Beyond two components, this can become a tedious and inefficient process, and the more popular approach is to solve for each component by the use of matrix algebra.

The multicomponent situation for a measured absorbance (normally located at a peak) can be summarized in general terms as:

$$A = k_1 c_1 + k_2 c_2 + k_3 c_3 + \ldots k_n c_n$$

This can be expanded to all components in a system as:

$$A_1 = k_{11} c_1 + k_{12} c_2 + k_{13} c_3 + \ldots k_{1n} c_n$$
$$A_2 = k_{21} c_1 + k_{22} c_2 + k_{23} c_3 + \ldots k_{2n} c_n$$

$$A_m = k_{m1} c_1 + k_{m2} c_2 + k_{m3} c_3 + \ldots k_{mn} c_n$$

In matrix terms the expressions can be rewritten as:

$$
\begin{vmatrix} A_1 \\ A_2 \\ \\ A_m \end{vmatrix}
=
\begin{vmatrix} k_{11}\, k_{12}\, k_{13} \ldots .k_{1n} \\ k_{21}\, k_{22}\, k_{23} \ldots .k_{2n} \\ \\ k_{m1}\, k_{m2}\, k_{m3} \ldots k_{mn} \end{vmatrix}
\begin{vmatrix} c_1 \\ c_2 \\ \\ c_n \end{vmatrix}
$$

or in shortened form as:

$$A = KC$$

This matrix notation represents the *K*-matrix approach to multi-component analysis. The calculation of an unknown is a two step process involving a calibration step and a component concentration determination step. These two steps are represented by two rearranged forms of the matrix equation:

$$K = A C^T (C C^T)^{-1} \text{ (Calibration)}$$

Where *A* and *C* are the absorption and
concentration matrices for the standards

$$C = (K^T K)^{-1} K^T A \text{ (Component concentrations)}$$

Where *C* is the calculated concentration matrix and
A is the absorbance matrix of the 'unknown' sample

An alternate matrix representation that is sometimes preferred is the *P*-matrix, which is expressed as:

$$P = C A^T (A A^T)^{-1} \text{ (Calibration)}$$

where *c*, the component concentrations, can be expressed directly as:

$$C = PA \text{ (Component concentrations)}$$

The advantage of the *P*-matrix over the *K*-matrix approach is that there is only one matrix inversion required by the process, and the unknown concentration matrix is obtained by a simple matrix multiplication of the *P*-matrix by the absorbance matrix of the unknown.

The normal implementation of the *P*-matrix and *K*-matrix methods is based on peaks, or selected absorbances, defined for all of the components being analysed. There are constraints imposed by these methods on the minimum number of standards required for the analysis. The number of standards required must be equal to or greater than the number of components to be analysed. In the case of the *P*-matrix, one more than the number of components is required as the minimum (*i.e. n* +1 standards). It is possible to use more than the minimum number of standards or more than the minimum number of analytical frequencies imposed by the number of components. In both cases, the process of using more than the minimum number required is known as over-determination—*i.e.* either overdetermination in standards or overdetermination in frequencies. Overdetermination can increase the precision of the measurement, and can help to compensate for errors caused by interaction effects.

It is possible to extend overdetermination to an extreme case where complete spectral regions are used rather than discrete frequencies. This approach is often known as a least squares curve fitting method. The advantages of this method over the point-based measurements are obvious—it removes the need to be specific and to select an analytical band, which removes the chance of a wrong or unsuitable band selection, and the use of a large number of data points reduces potential errors caused by noise.

One final comment on multi-component methods of calculation for the analysis of complex systems. The selection and preparation of standards is even more critical than in the simple two-dimensional case. Theoretically, the system can be standardized by the individual components without the need for the preparation of standards. However, with the exception of the analysis of gas mixtures, the preparation of standard mixtures is preferred over the use of pure component spectra. There are several reasons why mixtures are preferred when analysing condensed phase materials. Many samples are mixtures of solid and liquid phase materials and the inherent problems of analysing in the solid phase would make the use of pure standards impractical. Also, virtually all materials, when mixed in the condensed phase, exhibit some degree of mutual interaction. It would be impossible to represent the interaction effects if the materials were not analysed as mixtures. The standard mixtures must be a good representation of the final mixture to be analysed. That is, the concentration ranges of each component in the standard should bracket the expected concentrations (concentration ranges) for each component in a sample. Each standard mixture must be a unique mixture because the matrix process does not tolerate redundant (equivalent or duplicate) standards. It is not acceptable to make a single standard and produce additional standards by serial dilution of the single standard. It is good practice to produce more than the required number of standards, and to use some of the additional standards as a validation set.

7 SOURCES OF ERROR

Throughout this text, there have been several references to possible sources of error. In this short section, the common sources of error will be reviewed briefly in relation to the overall method of analysis, whether it is for a simple or a complex chemical system. The following are the most important sources of error:

Sample/Standards Preparation—it is important that the standards are representative of the sample. That is the expected concentration of the component or components are bracketed by the concentration ranges of the component(s) within the standards.

Sample Handling—the sampling method must be reproducible, and must be capable of supporting all forms of the sample, as well as the standards.

Poor Choice of Standards—emphasis has been placed on the quality of standards. Time is always well spent in ensuring that the standards are prepared with care. As noted, care must be taken to ensure that there are no redundant standards.

Solvent Effects—care must be taken to ensure that the solvent does not cause any severe distortions to the standard or sample spectra. Also, there must be no exchange effects caused by chemical interaction between the solvent (also applies to alkali halides used for solid sample preparation) and the sample.

Instrumental Errors—the instrument must be used with the same set of acquisition parameters for standards and samples. Any instrumental artifacts, such as fringes or spikes, must be absent from calibration standards, as well as the samples. Also, ideally, the spectra must be acquired with the highest practical signal-to-noise ratio. Note the word practical, because there are always time constraints.

Data Errors—numerical precision must be maintained across all samples. Numerical truncation, and distortions caused by numerical processing must not occur at any stage.

Poor selection of method—a common problem is that the person developing the method has a pre-conceived idea of the optimum method. This is not necessarily the most suitable or even the optimum method. It is important to be open-minded, and to test the method well with a validation set.

Incorrect or Unsuitable Method of Calibration—calibration may seem to be a simple exercise. However, it is easy to underestimate the system, and inadequately represent the sample under all sampling/experimental conditions.

Insufficient Understanding of Chemical System—this final point is often the fundamental reason for the failure of a quantitative method, and is also often the cause of many of the failure modes described above. It is easy to underestimate the magnitude of interaction effects between ingredients, the sample matrix or the sampling matrix, or a solvent (if added). It is also very important to study the spectra of the individual components, as well as the spectra of the components in the final sample.

8   CONCLUSION

This text has attempted to present a practical approach to quantitative analysis. The discussion has been limited to the traditional methods of calculation. These methods are quite adequate for a wide range of sample types. In recent years, new chemometric methods of analysis, such as Principal Component Regression (PCR) and Partial Least Squares (PLS) have been developed. These newer techniques provide a method for correlating traditional chemical or physical test methods for materials—such as viscosity, refractive index, and bromine number—with spectrometric response. A good review of these newer methods, as well as the existing methods, is provided in 'Computerized Quantitative Infrared Analysis', an ASTM publication edited by G.L. McClure. Quantitative analysis is of course well established for UV-visible and near infrared spectroscopy. Quantitative analysis applications with mid-infrared measurements offer many opportunities for industrial applications. With modern sampling techniques it is now practical to perform good quantitative infrared analysis with virtually every form of sample. Typically, success is limited only by time and imagination.

9   BIBLIOGRAPHY

'Computerized Quantitative Infrared Analysis', ed. Gregory L. McClure, American Society for Testing and Materials (STP 934), Philadelphia, 1987.

CHAPTER 8

Spectroscopy and Process Control

R.L. TRANTER

1 INTRODUCTION

Spectroscopic methods of analysis have been available for many years. Their fortunes have fluctuated and their main areas of impact have altered, but spectroscopic quantitative and qualitative analyses have continued to be developed. Outside the analytical support for research chemists, which has been dominated by NMR spectroscopy for the last 15 years, interest in industrial molecular spectroscopic analyses took a nose dive during the late seventies and early eighties as liquid chromatographic methods became the fashion. Now, this trend is being reversed. This is due partly to improved instrumentation and computing but also to a change in attitude as to the reason for carrying out an analysis in the first place.

Within the pharmaceutical industry, this is particularly clear. The vast majority of the manufacturing processes making the final pharmaceutical preparations are batch processes. Because of this, and the need to ensure a safe product, considerable attention has been focused on the quality of the final product with many different assays being carried out. Although the product is made by a validated process, less attention was paid to the control requirements of the process with the result that the final product analysis would occasionally reject batches of material. Some of these could be recovered, at a cost, but many were destined for the incinerator. Increasing competition in pharmaceuticals and keenly negotiated pricing policies with bodies such as the United Kingdom National Health Service has meant that such costs have had to be reduced.

An effective way of achieving this has been to focus on the intermediate stages of the manufacturing process and to make the small corrections at these stages to ensure that a consistent high quality product comes out at the end. This approach has created its own analytical problems; the traditional approach of taking samples and transferring them to a laboratory for analysis just cannot respond in the timescale needed to control the process. The assays must be carried out at the process and, preferably, they must be carried by the process operators who need to know the results. These constraints demand analytical methods which are robust, reliable, and responsive to the timescale of the process. The combination of automation and spectroscopy is the only way in which these objectives can be met.

This chapter will look at the spectroscopic aspects of the solutions; the automation will be mentioned only in passing when appropriate. Examples are taken from the UV, visible, near infrared, and mid-infrared spectral regions and from the raw material, intermediate, and final product stages of the process.

2   RAW MATERIAL CONTROL

There is a well known computer phrase: GARBAGE IN, GARBAGE OUT. This phrase applies equally well to manufacturing processes. If the raw materials going into a process are not right in either quality or quantity then a good quality product cannot result.

Ideally, a raw material needs to be identified and then assessed against the requirements of the process. The first of these causes few conceptual problems, but the second can be quite foreign to analysts who are used to quantifying chemical properties. For most of the processes of concern here, chemical properties are usually of secondary importance — it is physical properties such as hardness, particle size, or flow characteristics that determine how well the process performs.

Near-infrared spectroscopy offers much to help with these problems. Although spectral measurements can be made reasonably quickly and the instrumentation is robust, its main benefit comes from the complete lack of any sample preparation, apart from actually putting the sample into the instrument. When coupled with appropriate computation, this allows a robust raw material identification and assessment to be made in 4–5 minutes, the time taken to get the next sample ready for analysis. Two factors in the software are important in allowing this; control of the instrumentation and operator and the use of chemometrics.

2.1 NIR System Control

This is vital and is common to all the methods which will be described. The importance of control will be described in general terms to avoid repetition. If analytical methods are to be operated (but not developed) by process operators then it has to be recognized that they do not have the background knowledge to assess the performance of the instrument they are using (it can be argued that this is the case with many trained analysts!). Consequently, it is vital that the system forces calibration checks to be carried out, assesses the results of the checks and controls the access of the operator to an appropriate level of analytical control. In general, the latter is simply the option to run a predefined analytical method or to quit the system. The software monitors the instrumentation and the results of the assay and if it finds anything unusual it prints appropriate messages and directs the operator to a supervisor. If necessary they will not allow any further operation until the fault has been corrected.

The calibration checks need to be thought out carefully. In the case of the near infrared system they amount to a daily check of wavelength accuracy and a monthly check of absorbance performance. Experience has shown that this frequency is quite adequate, providing the instrument is left switched on all the time and an instrument engineer has not carried out any service work on the optical system. With other instruments it is necessary to carry out more frequent checks: for example, in UV measurements we are checking the spectral performance, essentially, every time a measurement is made.

2.2 Discriminant Analysis

In order to carry out a raw material identification, the spectrum of the sample must be identified. Classical methods of spectral interpretation or library searching are quite inefficient at this and do not have the required level of robustness. We do have one major advantage with this particular identification problem: we know what all of the raw materials are and by definition anything which is not in this set is a mistake. All that remains is to carry out an effective comparison of the sample spectrum with example spectra of all expected raw materials. The techniques of rapid discriminant analysis followed by full spectrum comparison are used to give a high degree of reliability.

Discriminant analysis uses only part of the information contained in a spectrum to carry out a very rapid comparison of the unknown against sets of five spectra of each raw material. No information is given about possible identity to 'aid' the process — it is quite unbiased. For this to occur, a library of reference spectra needs to be set up and calibrated. At the simplest level, consider the set of spectra given in Figure 1. Select any two wavelengths and then use absorbance of these wavelengths as the scales of a two dimensional x-y plot. Finally, mark on this plot the positions of each sample, Figure 2. It is seen that instead of a mass of overlapping spectra, a set of quite discreet points appears. Different batches of a material will appear as scatter about one point, as indicated by the group A in Figure 2. This scatter arises from spectral noise, variations in sample presentation, and batch to batch variations and it can be used to calculate a standard deviation contour for the group. If the wavelengths are chosen such that no group is within, say 10 standard deviations of another, then we have a very efficient way of determining whether or not the unknown is a recognized raw material (it will be more than, say, 4 standard deviation away from any group if it is not) and if it is which group it belongs to. In practice up to a dozen wavelengths may be required to pull the groups apart, but the principle is the same. The standard deviation contour is given the name of Mahalanobis unit.

To make the computation of the standard deviation contours simpler, it is assumed that all groups have the same shape and size. This is a reasonably

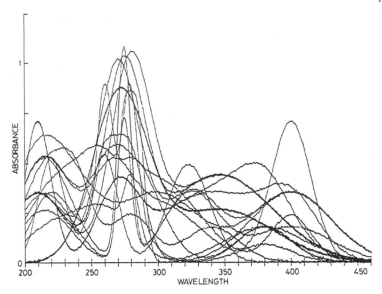

Figure 1   *Library of reference spectra*

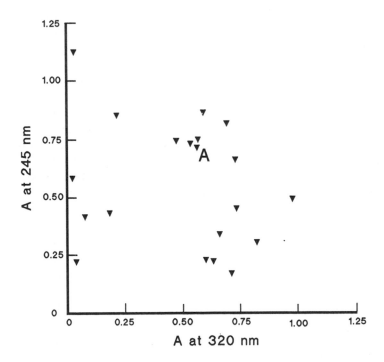

Figure 2   *Discriminant plot of library spectra using wavelengths* 245 nm *and* 320 nm. *Group A represents typical sample to sample variation*

good approximation but it does fall down occasionally and so it needs checking from time to time. Should many of the groups be found to have consistently different shapes then each group shape can be independently modelled using either a modified Mahalanobis procedure or by the SIMCA method (which is not described further here).

Although discriminant analysis is described here as a mechanism of library searching it is assuming a much more prominent role in all of our spectral analyses, including quantitative analysis. Examples of this will be given later.

Discriminant analysis is efficient but it is not entirely foolproof because it uses just a small proportion of the information available. For this reason it is backed up with a full spectral comparison. In this we form the sum of squares of differences in absorbance (or first derivative of asborbance) at each wavelength of the sample spectrum and the mean of the probable library group. This sum of squares can be regarded as a variance and so we can apply a variance ratio test on it and the variance calculated from the difference between the members of the group and the group mean. If the ratio is less than the tabulated figure at a given probability level then we have a high degree of confidence in saying that the unknown and the group mean are identical. These two tests are independent; the first is testing the difference between mean values and the second is testing variation about a mean. Satisfying both criteria gives a high confidence in the result.

2.3 Calibration Sets

An extension of this technique is to correlate spectral features with quantitative properties important for that raw material in its process. The problem here is identifying which properties are important! As already mentioned, physical properties are more likely to be important to the manufacturing process than chemical properties. Already some correlations of relevant properties with near infrared spectra have been obtained. Other, more relevant ones are being sought. This particular process simply identifies a small set of wavelengths whose absorbances correlate well with the property in question. Here we find another major problem which is common to all quantitative methods; that of calibrating the model.

Many of the materials measured are either mixtures or they have undergone some physical modification, such as grinding or granulation, or both. In each case it is extremely difficult to replicate these processes in the development laboratory to form a calibration set of materials identical with the material produced by full-scale plant. The consequence is that plant material has to be sampled, analysed by some reference method, and then used in the calibration model. Inevitably, the manufacturing process is working well when these calibration samples are taken and so the set has a very narrow range compared with the process control action limits. This has to be fully understood, because the set may be calibrating a non-linear model. Extrapolation outside the set is extremely dangerous. It also has to be recognized that the errors in the calibration model cannot be less than the errors in the reference analytical method. It is vital that good care is taken over the reference analyses.

Having calibrated the quantitative model it is then possible to use it to predict the result for the sample. As several models can be quantified using different wavelength sets, several analyses can be carried out simultaneously on the one sample spectrum. This gives an enormous time advantage, as the additional time for each analysis is measured in seconds rather than hours. It can be seen also that as the preceding discriminant analysis identifies the material, this can be used to select the specific set of quantitative analyses to be carried out for this material. If the material is not identified then no, possibly misleading, quantitations are made.

A major limitation of near infrared spectroscopy is its relatively poor sensitivity. A rule of thumb guide suggests a typical limit of detection of 0.1–0.5%w/w for quantitation in solids. This is usually satisfactory, but some potential contaminants may be at lower levels. It turns out that discriminant analysis and spectral comparison may be more sensitive to spectral changes and these can be used in a limit test mode where quantitative models are unreliable. Beyond this level of sensitivity it has to be asked if the reference method would detect the contaminant under operational conditions. If it would then near infrared spectroscopy should not be used for that part of the test.

3   FINAL PRODUCT ANALYSIS

Having looked at the beginning of the process we will now look at the assay of a finished product. It is necessary in this case as the product is an ointment and the two active ingredients are blended into the base as the last manufacturing stage before filling into tubes. It is also necessary because one of the active ingredients is an antibiotic which requires up to seven days to assay by classical methods.

3.1   FTIR Sample Presentation

Mid-infrared spectroscopy is appropriate to this analytical problem. First an ointment has a composition very similar to a Nujol mull so sample preparation can be reduced to spreading the ointment between potassium bromide plates. Secondly, the concentrations of the active ingredient give measurable signals away from the main paraffin bands, unlike NIR where the signals are not detectable. The third advantage is the availability of Fourier transform instruments. The advantages of FTIR are well known and have been commented upon several times in this book, but the three important ones, from the point of view of process control, are the speed of measurement, highly controlled instrumentation with wavelength referencing on every scan, and the availability of good computer control that can make the operation simple for an operator. Ointments are notoriously difficult materials to make and to analyse. These difficulties are most apparent when forming the calibration set of samples; production materials must be used and extraordinary care must be taken in the sampling and classical analysis to ensure results good enough to use as calibration values. The third difficulty is that the spectral features due to the active ingredients overlap each other to some extent, but each is severely overlapped by small features from the ointment base. These cause severe problems to simple multiple linear regression algorithms using absorbance spectra.

3.2 Principal Components Regression

Satisfactory results can be obtained only by going to principal components regression. This technique starts by making no assumptions about the components causing spectral variations: standard multiple linear regression requires the analyst to identify the known sources of variation and to then model each, using pure samples of each variation as a reference. In an ointment, it is not possible to identify all of the sources of variation. Clearly, the active ingredients and the ointment base are three components, but our work has shown that there are at least three more which have not been explicitly identified.

The first step in the analysis is to gather a calibration set of samples, roughly 50, which show the widest range of values for the active ingredients and which cover several manufacturing batches. The spectra of these are then subjected to a principal components analysis. This technique forms linear combinations of the original spectra in such a way as to maximize the contribution of each linear combination to the variation in the original data, without explicitly identifying the source of the variation. This process is most efficiently carried out using eigenvalue/eigenvector techniques and leads to orthogonal linear combinations variously called factors or principal components.

The process is shown pictorially in Figures 3 and 4. Figure 3 represents the spectra of five samples of varying composition. In Figure 4, the absorbance values at 20 different wavelengths in the spectra are plotted for just two of the samples. If the two samples contained only one component then the points would fall exactly on a straight line. Next the line F1 is drawn such that when all of the points are projected onto this line they show the maximum range of spread. Line F2 is then drawn. It will be at right angles to F1 and it will have the next biggest range of spread when that accounted for by F1 is removed. Between them F1 and F2 account for all of the variation in the data of Figure 4 assuming that only two components are present. It can be seen that they are linear combinations of the original axes and they are orthogonal; they are the first two principal components or factors. The coefficients of the linear combinations are the eigenvectors and the amounts of variation (the lengths of the lines) are the eigenvalues.

In reality, many more than two axes will be used and so many factors will be calculated. However, a large proportion of these will have eigenvalues very similar in magnitude to the noise in the original data and consequently they cannot be said to be significant and so may be discarded. Usually, about half a dozen significant factors are left. These factors can be applied to the original data to transform this data into so called 'factor spectra'. Factor spectra have

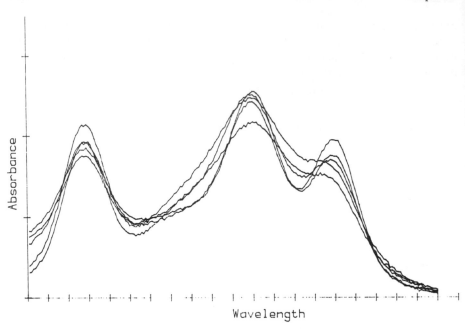

Figure 3 *Simulated spectra of five samples containing different amounts of three components*

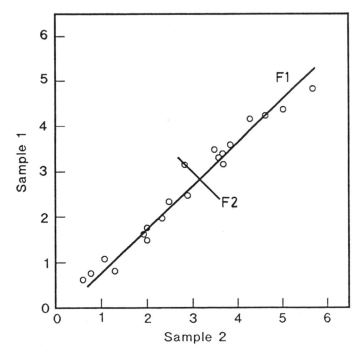

Figure 4 *Plots of absorbance in samples 1 and 2 for 20 wavelengths taken throughout the spectra in Figure 3. Lines F1 and F2 correspond to the first two principal components*

all the properties of the factors, particularly orthogonality, and this makes them very useful in modelling all of the significant sources of variation in the data, even those that were not suspected as being present. Correlations between the factor spectra and the calibration analytical values can now be sought just as in simple multiple linear regression techniques. The calibration equations can then be applied to a sample spectrum, transferred into the factor space to determine, in this case, the amounts of active ingredients. In our example, the recoveries and precisions of assay are very good and are probably limited by the quality of the reference analytical procedures. Like all other regression methods, principal components regression is dependent on seeing all expected sources of variation in the calibration set, if it does not it can fail. Failure has occurred in an example in which calibration was carried out using samples prepared in a small scale pilot plant. This information was used to predict the composition of samples on a full scale manufacturing plant, but this led to failure because other variables were not taken into account. Understanding this type of problem is critical to developing robust and reliable analytical methods.

4   INTERMEDIATE STAGE ASSAY 1

The first two examples have illustrated the particularly powerful combination of spectroscopy and chemometrics with much of this power coming from lack of sample preparation and good instrument control. When assays are considered using ultraviolet spectroscopy, sample preparation is almost inevitable as these measurements are best made on a solution of the sample. Because of the sensitivity of the UV measurement, the solutions must be very dilute $10^{-1}M$ to $10^{-6}M$. At these concentrations a particular problem is often caused by contamination of the analytical solution by careless handling. If the technique is to be used in a production environment then the system must be protected from the operator, and automated sample preparation is found to be the answer to the problem.

4.1   Automated Ultraviolet Assay

An example of this type of assay is a granulated powder mix where the homogeneity of the mixing needs to be determined before further use. An existing UV method was perfectly satisfactory but was simply in the wrong place in the laboratory. Sample preparation is quite simple, some 60g of sample dissolved in 400ml water followed by serial dilution to the analytical concentration. All of these stages needed to be automated and they were using a comprehensive in-house analytical systems automation language. The principal features of the automated method are that the operator is required to measure the volume of water and the amount of powder fairly crudely. The exact amounts are measured gravimetrically using a balance controlled by the system. Once the operator has carried out these operations and transferred the beaker to the magnetic stirrer everything else is automatic using a simple cartesian robot to perform the dilutions and a diode array spectrometer to make the measurements. The system also makes the required measurement of pH, giving two assays for the price of one. This is a feature of automated analytical systems; they can combine several assays in one system providing the sample goes down a concentration gradient through the system.

The focus here, though, is on the spectroscopy and not the automation. The diode array spectrometer was chosen for several reasons all of which are equally important. Diode array UV instruments are inherently robust as they have no moving optical parts, thus the wavelength scale shows no drift with time. This conclusion is reached from nearly nine years worth of data generated by the regular, frequent calibration checks on one instrument. The instrument gives a full spectral scan very rapidly, this is essential if the power of chemometrics is to be brought to the analysis. The technique of measurement gives an estimate of the error in each absorbance value at each wavelength. Such an estimate is unique in UV measurements and it allows much more robust statistical methods to be used as well as allowing more meaningful estimates of errors in the results. The instrument examines the raw data generating the spectrum on each scan and reports anomalous measurements. It also monitors other functions critical to reliable measurements and can give a status report on demand. Finally, all of the functions of the instrument are under the control of the automation system software, so the system can make its own checks on the instrument as well as direct it to make the correct sequence of measurements.

4.2   Multicomponent Analysis

A feature of diode array instruments that comes in for frequent discussion is its fixed resolution at about 2 nm equivalent bandwidth. This sort of resolution will introduce a small degree of distortion into most organic UV spectra, more in the fine structure regions of aromatic molecules. However, the resolution and the wavelength scale are fixed, so the distortion is constant. All of the calculation routines depend solely on a fixed spectral shape for each molecule, they have no knowledge of whether or not it is a true shape. This is not relevant. It is only necessary that the shape is constant.

Typically, quantitation of UV spectra will be done at a single wavelength for a single component system, or at many wavelengths for a multicomponent system. We have implemented the Hewlett–Packard maximum likelihood regression on our system. This is essentially a weighted least squares method making use of the absorbance error estimates. Alternative calculation methods are available such as principal components regression and a Kalman Filter multicomponent method. The latter is useful in a number of respects but particularly because it can show a degree of extra robustness to unexpected or missing components when compared with ordinary linear regression.

All of these calculation methods again highlight the problems of calibration. It is particularly important in UV calibration to understand these as calibration solutions of reference substances can be made readily. If care is not taken in the experimental design for the calibration set to look for, and account for, interactions between the analytes and solvent, pH effects, ionic strength effects, and temperature effects then quite serious errors of the believable but wrong variety can result. Figure 5 illustrates the effects of both pH and temperature on one system we have examined. It was concluded from this work that pH needed to be controlled to better than 0.1 pH unit and temperature to better than ±0.1°C. Neither of these conditions are usually applied to the measurement of UV spectra for routine analysis. To account for possible deviations from the calibration, we are now introducing a form of discriminant analysis on the sample before it is analysed quantitatively.

5 INTERMEDIATE STAGE ASSAY II

The final example is of colour measurement in the visible part of the spectrum. Here the problem was to improve the effectiveness of measuring the clean-up stages of a process intermediate and, hence, to optimize the recovery of the material. The material is in a liquid stream and is highly coloured so no sample preparation is required. The current visual estimation of colour is subjective and prone to operator effects, such as tiredness, lighting conditions, and who is making the measurement.

5.1 Colour Assays and Discrimination

The instrument chosen for this work was developed over several years by a company in collaboration with ourselves. As one might expect by now it is robust and controllable. It also makes measurements very rapidly, up to 30 per second, although this is not a requirement in this application — once a minute is fast enough.

The measurements are converted into the CIE tristimulus coordinates which may then be plotted in the standard CIE colour space diagram for an estimate of colour (Figure 6). However, these coordinates are not user friendly so we transform them into the CIE polar coordinate system in which the two principal parameters P and H are easily interpreted into a colour. H is the angle of the sample from the reference line. As this line is red, increasing angles mean that the colour is moving through orange, yellow, green, and into blue (Figure 6). The P value gives the depth of colour. Small values mean the sample is close to the equal energy point E and, hence, are pale colours, large values mean strong colours. H is a parameter that essentially depends on the molecular type while P depends on its concentration. These two can be used to discriminate between quite similar molecules over a range of concentrations, including concentrations which block all light reaching the blue detectors (Figure 7). Such discrimination then allows a meaningful monitoring of the process stream. Not only can the target molecule be monitored but so can impurities from the process. The final point about this system is the necessary calibration check. At present this is limited to the instrument's response to white (no colour) and black (no light), as there are no recognized colour transmission standards

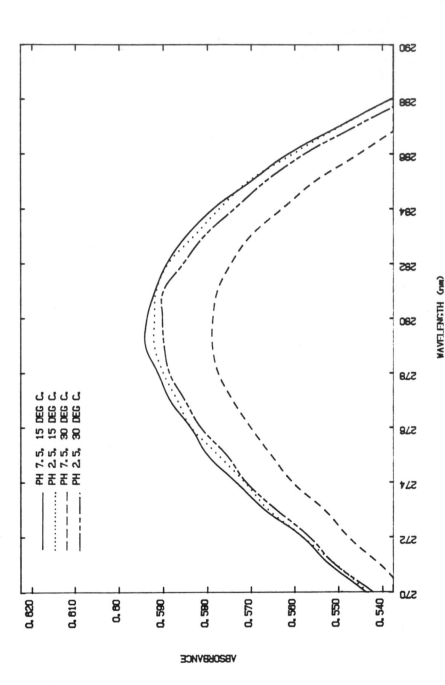

Figure 5 *Effect of changes of temperature and pH on the spectrum of a sample. The small absorbance and wavelength shifts shown are enough to invalidate a multicomponent*

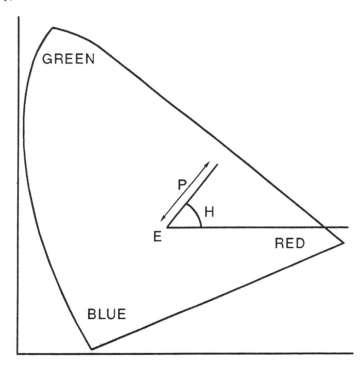

Figure 6 *The CIE polar colour scale*

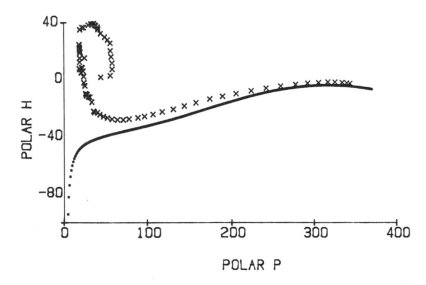

Figure 7 *Plot of colour for two samples over a wide concentration range. Solid line is the pure reference substance, the crossed line is a process sample*

available. Our experience over a couple of years has been that proper use of these is adequate for most purposes. However, standard colours are useful for checking the performance of individual detectors in the instrument and we are making some headway in defining what these should be.

6 CONCLUSIONS

Quality control is as much a part of the state of mind as it is part of analytical chemistry. It is most effectively and efficiently implemented through proper process control measurements rather than by exhaustive post-mortem analysis. But, if this is to be the case the measurement systems must be highly reliable and robust. Spectroscopic systems have many characteristics that allow them to be used as process measurement systems, particularly when they are combined with chemometrics methods. However, it is necessary to meet several criteria:

   (i)   the right instrument must be chosen both for its spectroscopic range and for its construction;

  (ii)   the performance of the instrument must be checked frequently, thoroughly, and automatically by the system;

 (iii)   the complete system, including the operator interaction, must be under control and this is best achieved with automated systems;

 (iv)   the assay must be properly calibrated using effective experimental designs and a good understanding of the limitations of the chosen model and procedure;

  (v)   the user interface must be kept as simple as possible.

7 BIBLIOGRAPHY

Discriminant Analysis
H.L. Mark and D. Tunnell, *Anal. Chem.*, 1985, **57**, 1449.
'Multivariate Statistical Methods: A Primer'. B.F.J. Manly, Chapman and Hall, London, 1986.
'Chemometrics: A Textbook', D.L. Massart *et al.*, Elsevier, Amsterdam, 1988.
'Chemical Pattern Recognition', O. Strouf, Research Studies Press Ltd., John Wiley and Sons Inc., Colchester, 1986.
Regression Methods
'Applied Regression Methods', 2nd Edition, N. Draper and H. Smith, John Wiley and Sons Inc., Colchester, 1981.
P.M. Fredericks, J.B. Lee, P.R. Osborn, Dom A.J. Swinkels, *Appl. Spectrosc.*, 1985, **39**, 311.
H.N.J. Poulisse and P. Engelen, *Anal. Lett.* 1980, **13** (**A14**), 1211.
S.C. Rutan and S.D. Brown, *Anal. Chem. Acta*, 1984, **160**, 99.
Diode Array Instrumentation
'The Diode Array Advantage in UV/VIS Spectroscopy', A.J. Owen, Hewlett–Packard, 1988, HP Publication No. 12–5954–8912 (This booklet contains an extensive bibliography.)

CHAPTER 9

FTIR in Quantitative Analysis and Automation

P.S. WILSON

1 INTRODUCTION

FTIR: The advent of Fourier Transform Infrared spectroscopy has greatly increased the use of infrared spectroscopy. The advantages of FTIR are well known.

(i)   Higher Throughput; the infrared power reaching the sample from the sources is considerably greater than with a dispersive system. This is true even at high resolutions.

(ii) Signal/Noise Ratio; the FTIR spectrometer has an inherently higher signal to noise ratio than a dispersive infrared system. This enables techniques such as microscopy and GC–IR to be used with infrared.

(iii) Speed; perhaps the most important advantage of FTIR is the speed with which data can be accumulated. A complete spectrum is collected to a resolution of 4 cm^{-1}, in less than a second. In a minute sixty four spectra can be collected, and averaged to produce a high quality spectrum.

The speed of spectral acquisition also greatly extends the use of FTIR for mathematical manipulations. Frequency based subtraction is common place. However, since it is possible to collect twenty or thirty spectra an hour using FTIR, quantitative analysis becomes a much more attractive proposition.

FTIR has become increasingly the technique of choice for the quality control laboratory. The speed and improved signal/noise ratio has resulted in its use for qualitative and more recently quantitative analysis.

The increased use of FTIR in quantitative analysis has resulted in the development of several analytical packages;

MATRIX ANALYSIS

FULL MENU LEAST SQUARES FIT

PRINCIPAL COMPONENT REGRESSION ANALYSIS

As FTIR replaces the dispersive IR-technique so its value to process control becomes more significant. The quantitative techniques are of particular significance for the use of infrared in quality control. The quality control chemist needs a full spectral characterization of a material. However, often the presence of a minor component and its concentration must be quantified.

The various quantitative packages are described below. No one quantitative method is suitable for all applications. Each has its part to play dependent on the spectra to be obtained from a particular calibration series.

2   INTRODUCTION TO QUANTITATIVE ANALYSIS

Quantitative analysis is based on Beer's Law stating that the radiation absorbed by a substance is exponentially related to the concentration of the substance and the pathlength of the radiation through the substance. For a single component at a single frequency Beer's Law is expressed as;

$$A = abc$$

where A = the absorbance at a particular frequency, a = the absorbtivity for the absorbing species at that frequency, b = pathlength, c = the concentration of the species.

According to Beer's Law a plot of absorbance *versus* concentration should give a straight line.

Deviations may occur due to instrumental or chemical factors and result in curvature of the plot of concentration *versus* absorbance (Figure 1).

Chemical effects causing deviation from the norm are;
(i)   dissociation
(ii)   self interaction
(iii) complex formation
(iv)   solvolysis

Instrumental effects causing deviation are;
(i)   measurement of spectrum at insufficient resolution
(ii)   stray radiation
(iii) poor instrument performance

For multicomponent mixtures Beer's Law also applies. The absorbance is an additive property of all the absorbing species, *i.e.* total absorbance.

$$A = a b c + a_2 b c \ \ldots \ldots a_n b c_n$$

where a_n = absorbtivity for the nth absorbing species, c_n = concentration of the nth species, b = the constant pathlength for all components.

The Bio-Rad FTIR software offers several packages for quantitative analysis. The methods are tailored to the requirements of the user. They extend from simple menu operation to Principal Component Regression Analysis. The latter is an advanced package which reduces operator error to the absolute minimum. Each software package has its advantages according to the needs of the analyst. The packages are outlined below.

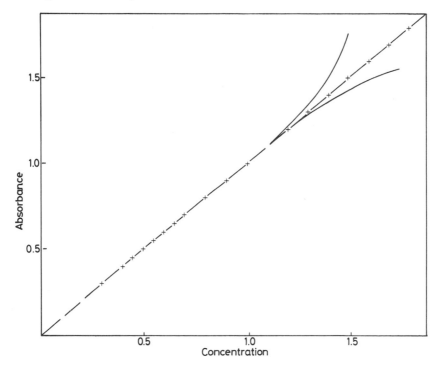

Figure 1 *Plot of Beer's Law* (A=kC)

2.1 Specfit

Specfit is a multicomponent quantitative analysis package which uses the full spectra. The calibration standards employed are mixtures.

The specfit analysis works by performing a least squares fit of an unknown spectrum to a group of calibration spectra. The components of the unknown spectrum are then estimated from the callibration spectra using the same linear model.

The simplicity of specfit is determined by the use of the complete spectrum. There is no pre-selection of peaks or baselines.

The results include a print-out of the concentration of various components, the weighting vectors used, the components response curve, residual spectra, and an estimate of the components pure spectra.

The residual spectrum can be used to show up impurities occurring in the unknown, but not in the calibration spectra.

Only spectra which have a linear relation with sample quantity can be employed using specfit such as;
(i)   Absorbance
(ii) Kubelka–Munk
(iii) Photoacoustic spectra
Transmittance spectra cannot be used.

Specfit quantitative analysis is carried out in three stages.

(i)   Collection of spectra for calibration.

(ii) Calibration.

(iii) Analysis of unknown and presentation of data.

Specfit is a completely menu-driven form of quantitative package. On entering specfit the menu page illustrated in Figure 2 is presented. From this page the analyst enters;

(1) The components, *e.g.* EtOH, IPA, H_2O, to be determined.

(2) The units of concentration, *e.g.* p.p.m., %, Molar.

(3) The definitive spectral data for the calibration spectra, *i.e.* Resolution and Spectral range.

(4) The weighting function given to each data point used for calibration. The significance of each weighting is shown in Table 1.

Once the primary data are entered the operator then selects the data path from which the data used in calibration and analysis are to be extracted. This is done through the parameter menus (Figure 3).

The calibration spectra may now be selected and the concentration of the various components registered (Figure 4).

SPECFIT-QUANT

MAIN CALIBRATION MENU

Property names: _____   _____   _____

_____   _____   _____

Property units: _____

Starting wavenumber: _____cm−1

End wavenumber: _____cm−1

Resolution: _____cm−1

Weighting function: None Threshold T−1   T**2

F1–Analyse menu   F6–Load/save calibration

F2–Display menu   F7–Perform calibration

F3–Calibrate menu   F8–Edit calibration script

F4–Parameter menu   F9–Load/save calibration script

F10–Run PRESS procedure

Figure 2   *The Calibration Menu*

Table 1 *The Weightings Available*

None   All data points are weighted the same. This weighting should be used initially before experimenting with alternative weightings.

Threshold Based on absorbance measurements, and uses two parameters "WEIGHTING THRESHOLD £1" and "WEIGHTING THRESHOLD £2" to determine the weighting. All spectral points between threshold £1 and threshold £2 in absorbance units get a weighting of 1; all spectral points less threshold1 or greater than threshold £2 are ignored.

T^{-1} The weighting at each point is the inverse of the transmittance value. Thus strong bands (low transmittance) are weighted more heavily than weakbands and the background is weighted least of all. This method is expected to be most useful when no strong non-linear bands are present, and a large proportion of uninteresting baseline exists in the spectra.

T^2 The weighting at each point is the square of the transmittance value. Strong bands (low transmittance) are weighted much less than weak bands. This method is most useful for rejecting data from highly absorbing bands which may be non-linear.

Combined   A combination of T^{-1} and T^{-2}, which attempts to use the best parts of both methods. It should emphasize moderately absorbing features at the expense of both baseline and non-linear bands.

SPECFIT–QUANT

PARAMETERS

Datapath:
/usr/IR1

Weighting threshold £1: _ _ _ _ _ _ _ _ _
Weighting threshold £2: _ _ _ _ _ _ _ _ _

F1: Analyse menu
F2: Display menu
F3: Calibrate menu
F4: Parameter

Figure 3 *The Parameters Menu*

Define Calibration Menu

Name of spectrum   £1   _ _ _ _ _ _ _ _

Property   £1:   _ _ _ _ _ _ _ _   (%)

Property   £2:   _ _ _ _ _ _ _ _   (%)

F1–Analyse menu   F7–Last calibration spectrum
F2–Display menu   F8–Next calibration spectrum
F3–Calibrate menu   F9–Delete this calibration entry
F4–Parameters menu   F10–Done entering spectrum

Figure 4   *The Calibration Menu*

Through this menu the operator enters the name of each calibration spectrum to be used and is subsequently prompted for the concentrations of the various components of the mixture defined by the spectrum.

When the last value is entered the procedure is exited and the user is returned to the page illustrated in Figure 2. From Figure 2 it can be seen that pressing a function key F7 performs the calibration from which the data can be calculated and stored.

The most important part of establishing a calibration is a verification step used after calculation of the calibration data. In Specfit the procedure is called PRESS and proceeds in the following steps:

(1) From the calibration set, one spectrum is removed.

(2) If the original set was 10 spectra a calibration is performed with nine spectra. The 10th spectrum is analysed as an unknown.

(3) The concentration of the spectrum calculated as an unknown is compared to the concentrations entered by an operator and a root mean square of the difference is calculated.

(4) A print-out is produced (Table 2).

In an ideal world the entered and calculated values should be identical. Spectra where recalculated values differ considerably should be eliminated from the calibrated.

The unknown spectra can then be entered on the page of the menu illustrated in Figure 5.

The spectra is entered, the analysis is executed and the concentrations of the components are displayed accordingly.

After analysis the unknown can be compared to the average calibration spectra (Figure 6). Plotted and manipulated for a further check on the accuracy of the analysis.

Table 2 *Print out of calibration*

name of cal. std being analysed	Component A	Component B	Component C	
00–03–07A.DT	0.000	0.3364	0.6636	—input concn.
	−0.0052	0.3386	0.6699	—back computed concn.
	−0.0052	0.0022	0.0063	—difference

0.0058 = root mean square of difference

SPECFIT–QUANT

ANALYSE MENU

Unknown spectrum: _ _ _ _ _ _ _ _(absorbance)
Property £1 _ _ _ _ _ _ _(%)

Property £2 _ _ _ _ _ _ _(%)

F1–Analyse menu   F6–Start new report
F2–Display menu   F7–Add results to report
F3–Calibrate menu
F4–Parameters menu

Figure 5 *The Analyse Menu*

2.2. Multicomponent Quantitative Analysis

In Specfit the complete data set is employed for quantitative analysis. However, it may be that not all the spectral data will be relevant to quantitative analysis. The second package offered by Bio-Rad is called Quant 32 and this package employs *P*- or *K*-matrix analysis.

Each method of matrix analysis has advantages and/or limitations. However, it is in the structure of the method that the Quant 32 package differs and it is this that will be concentrated on here. The procedure for Quant 32 is as follows;
(1) Collection of spectral data.

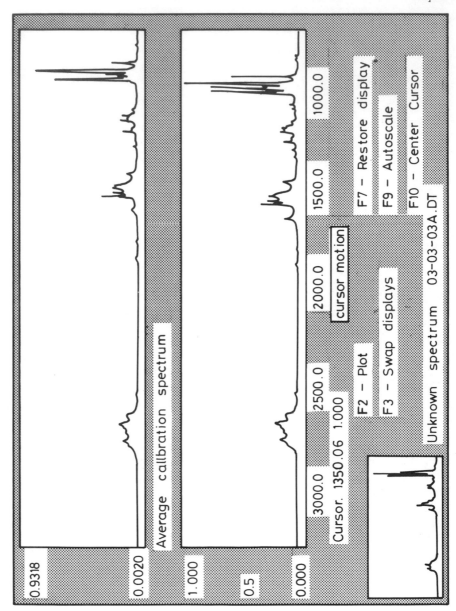

Figure 6   *The Specfit Display Field. The Specfit Display consists of two spectral displays, one above the other, and a radar box. Its roll/zoom and scroll operations are the same as the standard display function; except that the bottom display is fully interactive while the top display tracks the bottom display's X-axis motion while continuously autoscaling in the Y-axis mode. The radar box shows the spectrum displayed in the bottom window.*

(2) Display of spectral data.
(3) Selection of spectral data which seems to be most concentration dependent.
(4) Create a spectral reduction script for use in calibration.
(5) Calibration.
(6) Back calculation of entered concentrations.
(7) Analysis of unknowns.

Step 1   Is a formal stage and can be performed through a menu step or via a macro.

Step 2 and 3 In order to best identify the spectral detail related to concentration several methods may be employed.

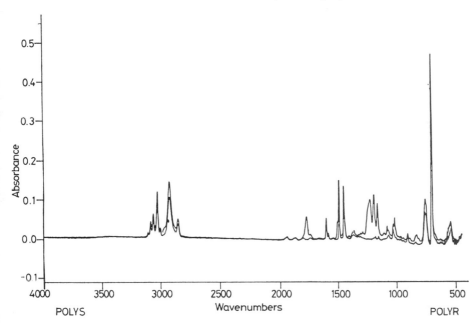

Figure 7   *Spectral overlay of two polystyrene types*

From each of these methods it should be possible to identify those spectral components which change with the concentration of minor substance components. These can be entered into a reduction script.

Step 4   Take a spectra shown in Figure 10. From steps 2 and 3 several peaks may be identified as changing with the concentration of components A, B, and C.

There are several ways in which the data from this spectra may be used;

(i) Peak area: the area of peak may be employed. The area is defined by a baseline drawn between two data points (Figure 11).

(ii) Peak Height: the height of a defined peak may be employed and can be defined as the maximum point within a spectral region.

(iii) Grid: a grid may be applied to a spectral region defined by a baseline drawn between two points (Figure 12).

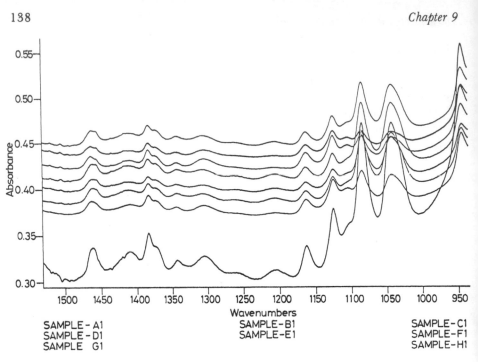

Figure 8 *Multiple Spectral Display: mixture of ammonium tartrate, ethylene glycol, and iso-propanol in water*

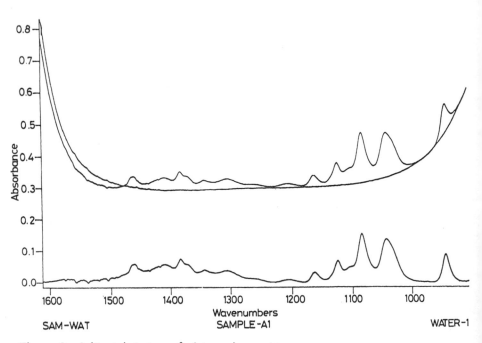

Figure 9 *Subtracted spectrum of mixture minus water*

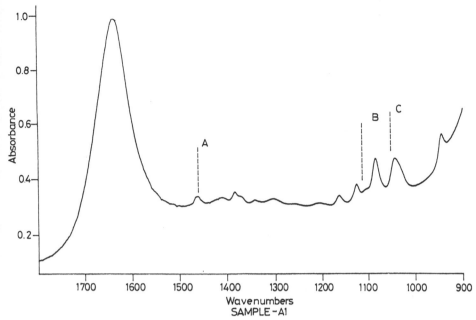

MIXTURE OF AMMONIUM TARTRATE, ETHYLENE GLYCOL AND ISO-PROPANOL IN WATER

Figure 10 *Spectrum showing mixture of ammonium tartarate, ethylene glycol, and iso-propanol in water*

The spectral data may be written into a file known as a reduced script,
   e.g. XYZ rs
         area 3000 – 2800 h (3010 – 2790)
         point 1760 – 1680 max
         grid (900 – 840 6) b (900 – 840)
is translated as; from the complete spectra use the following data;
(i) Measure the peak area between 3000 and 2800 cm⁻¹ defined by a baseline drawn between 3010 – 2790 cm⁻¹.
(ii) Measure the maximum peak found between 1760 and 1680 cm⁻¹ in terms of peak height.
(iii) Apply a grid to an area of 900 – 840 cm⁻¹ defined by a baseline drawn between the same two points. The grid means that the absorbance value will be measured six times across the defined region.

Step 5   Calibration
            The first stage of calibration is to establish the calibration parameters. This is done by entering the following parameters.
            (i) rscript = XYZ rs. Enter the rscript determined above.
            (ii) datapath = /usr/IRI where the data collected in Step 1 is stored.
            (iii) Propnames = the names of the components for which concentrations are to be found.
            (iv) Propunits = the units of concentration %, ppm, mM, *etc* . . .

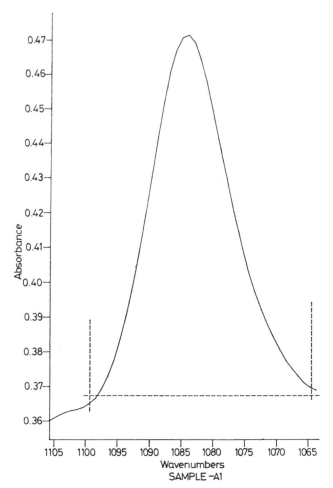

Figure 11   *Peak Area Taken from component B (Figure 10)*

The calibration can then be entered by typing the name of the spectra and the concentration of each component.

Step 6   Back calculation of Entered Concentrations
                When the calibration is completed, the following data is available.
                From Beer's Law.
                $A = a\,b\,c$

The absorbance is measured from the spectra, the concentration value has been entered, and the constant of proportionality for the whole series is calculated. Using this last value the concentration can be recalculated to produce a series such as that shown in Table 3.

In such a table the original concentration, the recalculated concentration, and the difference are shown. A table such as this can show two things.

(i) If all the results are at odds then it is most likely that the spectral data selected are incorrect and should be re-checked.

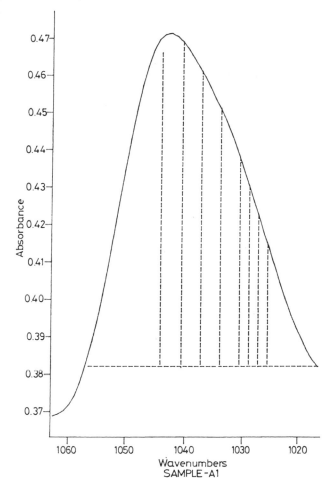

Figure 12 *Grid Taken from Component C (Figure 10)*

(ii) If a single value is wrong, then it is likely that the entered concentrations are wrong and the calibration standard should be made up again.

Step 7 Analysis of the unknown. The name of the spectrum is entered and the concentrations of the components required is calculated and printed out.

2.3 Principal Component Regression Analysis

Specfit is simple because it is menu-driven and because the total spectral data are used. However, there is still some operator error involved in getting the weighting factor correct.

Quant 32 can introduce operator error both in the selection of spectral data for a reduction script and in the preparation of standards.

Ideally a quantitative package should be set out to reduce the operator error

Table 3 *The report generated from the calibration*

Fri Nov 21 16:27:15 1986   algorithm=p   rscript-sugarl.rs

Sample	SUGAR-A WT%	SUGAR-B WT%	SUGAR-C WT%	
SUGAR1A	2.0720	1.8520	2.1240	(Input concn.)
	2.1130	1.9344	2.0724	(Back computed concn.)
	−0.0410	−0.0824	0.0516	(Difference)
SUGAR2A	4.2000	1.9860	3.0340	
	4.1657	2.0202	3.0233	
	0.0343	−0.0342	0.0107	
SUGAR3A	3.0140	3.9600	6.1040	
	3.0035	3.9320	6.1199	
	0.0105	0.0280	−0.0159	
SUGAR4A	3.0660	3.0320	2.1600	
	3.1997	3.1467	2.1910	
	−0.1337	−0.1147	−0.0310	
SUGAR5A	6.0200	2.0040	3.1380	
	5.9713	1.9126	3.1461	
	0.0487	0.0914	−0.0081	
SUGAR6A	1.0760	1.0880	1.0580	
	1.0282	1.0506	1.0060	
	0.0478	0.0374	0.0520	
SUGAR7A	3.0380	3.0820	3.0980	
	2.9658	2.9724	3.0957	
	0.0722	0.1096	0.0023	
SUGAR8A	2.0620	3.0560	2.0600	
	2.1154	3.1192	2.1094	
	−0.0534	−0.0632	−0.0494	
SUGAR9A	2.1620	5.0340	3.0460	
	2.1658	5.0373	3.0343	
	−0.0038	−0.0033	0.0117	
SUGAR10A	4.0660	1.1240	4.1520	
	4.1380	1.2004	4.1592	
	−0.0720	−0.0764	−0.0072	
SUGAR11A	3.1200	4.3600	3.1660	
	3.0634	4.3412	3.1111	
	0.0566	0.0188	0.0549	
SUGAR12A	1.1200	2.0760	1.0540	
	1.0911	1.9943	1.1252	
	0.0289	0.0817	−0.0712	
RMS errors	0.0597	0.0710	0.0382	(root mean square of difference)

to the absolute minimum. Without automation it is difficult to reduce errors in making up standards. However, it is possible to reduce error when treating spectral data. Principal component regression analysis is an advanced quantitative package which fulfils this requirement.

3 PCR AND PLS

3.1 An Introduction to PCR and PLS

Principal component regression (PCR)* and partial least squares (PLS) analyses are intended for routine, rapid and non-destructive determination of the chemical composition (or other properties of materials) by infrared spectroscopy.

The methods perform multicomponent analysis using full spectra. They are simple to use since they deal with spectra as a whole, thus, they require no selection of peaks or baselines. Since they work with complete spectra, the 3200 data system display capabilities may be used to show the results.

PCR and PLS are robust analytical methods. They may be used when the spectral bands to be analysed are overlapped or are affected by baseline variations, spectral noise, purge variations, sample impurities, and non-linear response. In addition, PCR and PLS analyses can detect errors due to poor sample preparation, sample impurities, and operator errors. This provides a warning that the analysis results may be incorrect.

PCR and PLS analyses must be user calibrated. Once the analysis has been calibrated,it may be used repeatedly to analyse samples of similar composition with little user interaction. Therefore, the routine analyses of samples may be carried out by a technique or a production worker.

The calibration requires preparation of a set of calibration standards which are mixtures of known composition for the components being analysed. The calibration of these standards is determined by gravimetric analysis, some other chemical analysis, or the measurement of a property of the material being analysed.

The number of components which can be simultaneously analysed by PCR or PLS is not fixed. The number of calibration standards is limited only by available computer memory. A greater number of calibration samples should exist than the number of absorbance sources present in the samples. Such sources include each known chemical species as well as any sources of spectral 'interferences' such as chemical interactions, physical effects including optical interferences, and baseline variations. The PCR/PLS analysis techniques ignore any source of absorbance that remains constant in all standards and samples. They also automatically compensate for sources of spectral interference when the interference is present in both the samples to be analysed and the calibration samples. Generally, using a greater number of calibration standards than the known components, improves the analysis by compensating for unanticipated sources of spectral variation in the analysis.

Concentration is the typical measurement, but other attributes, such as engineering properties, may be used instead of concentration. Viscosity is one

---

*The basic statistical methodology of principal component analysis is also known by several other names. In the chemical literature, it is frequently called 'factor analysis'. The terms eigenvector analysis, latent vector analysis, or singular value decomposition can also be used to describe this method.

example. The chemical (or chemicals) being analysed may be present in more than one species in solution, but specific knowledge of the species present in the solution is not needed. However, an adequate number of standards must be used to create the calibration. The number of standards to be included in the calibration increases with the complexity of the sample.

The success of an analysis depends on correctly designing the calibration experiment. The standards must be a true representation of the unknown samples.

3.2 An Overview of the PCR and PLS Approach

PCR and PLS are performed in three stages. The stages (as shown in Figure 13) are:
(1) Calibration
(2) Validation and optimization
(3) The analysis of the unknown samples

3.2.1 The Calibration Stage. In the calibration stage, the spectra are collected for a set of samples. The composition of these samples is measured by a reference method. A file is created which correlates the spectra to the known composition and the PCR or PLS calibration program is then run.

For efficiency and to insure valid results, the calibration set should be constructed by designed experiments, and the known concentration values in the calibration set should bracket the nominal values expected in the analysis of unknown samples.

3.2.2 The Validation and Optimization Stage. Once the calibration is done, the second stage is to do the cross-validation to test the validity of the calibration and to optimize it. This stage is frequently omitted when carrying out other quantitative IR analysis techniques, but is essential in the PCR/PLS approach.

Validation is carried out using another set of calibrated samples known compositions. This second set of samples is called the 'validation set'. It is used to optimize the prediction of the PCR/PLS model by choosing the 'Rank' of the model.

The ability of the PCR/PLS model to predict accurately the concentration of unknown samples depends on the rank of the analysis. The rank is the number of mathematical components which are useful in predicting the concentrations. Generally, this number will be less than the number of standards, if (as recommended above) a sufficiently large number of standards is used.

The rank is determined by comparing the PCR/PLS predictions for a range of ranks and comparing the results with the known values. A report is generated that includes a summary of the standard deviation, the bias (the mean deviation), and the standard deviation corrected for bias (the precision). A plot of standard error *versus* rank should be made, and the expected behaviour of the standard error is to have a minimum at the optimal rank. This optimum rank is then assigned to the calibration for the analysis of the unknown samples. In addition to determining the optimum rank of the analysis, the validation also

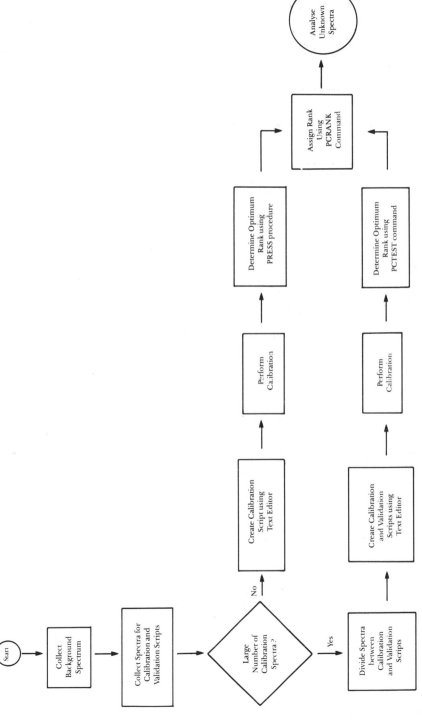

Figure 13 *A guide to the PCR and PLS approach*

shows the prediction errors to expect when using the analysis to determine the composition of unknown samples.

As an alternative when a separate validation set of samples is not available, the validation of the PCR and PLS model can be carried out using the 'PRESS' procedure. The PRESS procedure (Predictive Residual Error Sum of Squares) takes the calibration spectra one at a time and removes them from the calibration set. A calibration is performed on the remaining spectra, and the removed spectrum is analysed as an unknown.

However, depending on the number of spectra, number of components, size and resolution of spectra, *etc.* the PRESS procedure can take considerable time to run. When a large number of calibration spectra are available, the faster method is to divide the spectra between the calibration and validation sets.

3.2.3   The Analysis of Unknown Samples. The last stage is to analyse the unknown samples using the previously computed and optimized calibration. The choices relating to the analysis of the unknown samples are made during the first two stages, which makes the analysis of the unknown samples quite simple. This allows the analysis to be run from a simplified menu interface.

A useful feature of PCR and PLS analysis is that the 'spectral noise level' of the unknown samples can be measured and compared with the noise level in the calibration spectra. The comparison is called the 'F ratio'. This is an important part of the validation of the analysis results; if the F ratio for a sample is much higher than the F ratios observed during cross-validation, it is an indication that the sample is not similar to the calibration set being analysed and that the predicted concentrations may not be accurate.

4   APPLICATION OF QUANTITATIVE ANALYSIS

Quantitative Analysis of multi-component mixtures are not easily developed. Development takes a great deal of care and attention. However, it must be the aim of the operator that, having developed the method, the analysis can produce accurate results for as long as the method is required.

Once the analysis has been satisfactorily tested the ultimate goal is its application to on-line automation.

FTIR is relatively new to the chemistry quality control laboratory. In one area of industry it has been the technique of choice for quality control for over ten years. The semi-conductor industry has employed FTIR on the production line for a variety of characterization measurements. Particular amongst these have been film thickness measurements, analysis of carbon and oxygen content, and also for the quantitative analysis of boron and phosphorus in silicon passivated glasses. The application of FTIR to this analysis is particularly important, as silicon wafers with high phosphorus content have unstable com-

---

Note — the PCR/PLS methods provide two different validations:
      (1) the cross-validation of the calibration and
      (2) validation of the analysis of each unknown sample

positions and unreliable performance. The IR absorbance of boron and phosphorus oxides occur at approximately 1330 and 1400 cm⁻¹ respectively. The spectral bands are broad and when both constituents are present there is considerable overlap. This make it difficult to employ peak areas or peak height to quantitative analysis.

In these instances (Figure 14), where it is difficult to discriminate spectral information, Principal Components Regression Analysis is ideal.

The calibration set is tailored to meet the critical concentration range of boron and phosphorus in silicon wafers. High and low acceptance limits are set and these limits are used to drive an autofeed–autoscan (AFAS) device. The

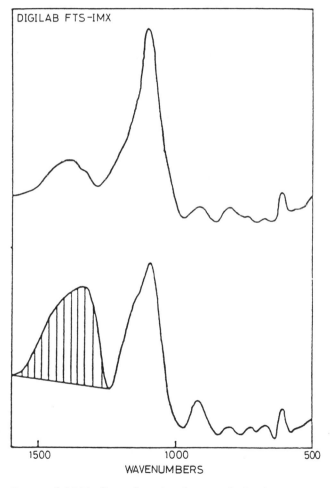

Figure 14 *Spectra of BPSG films of varying boron and phosphorus concentration. In the Multicomponent Analysis procedure absorbance measurements were made over the overlapping bands due to the boron and phosphorus oxides as shown in the lower spectrum*

AFAS system can accept up to 50 wafers and will sort these according to the present limits of boron and phosphorous.

Fibre optics is another area to which quantitative packages have been applied. The oil and food industries have particular interest in the application of fibre optics in the near infrared.

Quantitative analysis coupled with fibre optic probes enables the operator to sample a production line at regular intervals and determine the concentration of various components in a mixture. The concentration value can then be employed to trigger additions to a process in order to correct total concentration deficiencies.

The direction of FTIR is very much towards process control. All manufacturers are now producing stand alone key-pad operated optical benches.

In these systems data is collected into RAM with instant access to a printer.

However, it is worth noting that such is the power of FTIR, that a powerful data station is essential. Quantitative analysis in process control can tell the operator that something has gone awry. Access to the data enables the operator to diagnose the fault.

Exchange of Infrared Data using the JCAMP-DX Format

H. SOMBERG

1 INTRODUCTION

With the development of Fourier Transform Infrared Spectrometers and the increasing use of computers in analytical chemistry, the interest in exchanging spectral data between different institutions has also increased.

Nevertheless it took several years before an exchange format for infrared data was developed by the JCAMP committee (Joint Committee on Atomic and Molecular Physical Data) in the USA. During development draft versions of the format were distributed to several institutions for revision and test implementations. Meanwhile, a detailed description of the format had been published[1] and had been mainly accepted.

So why do we need this article?

There are still several hurdles. For further acceptance the format needs to become generally known, in other words it needs more 'publicity'.

The original paper describes the exchange format in great detail but is, in the author's opinion, intended for use by programmers as a reference. The description is sometimes difficult to read for the non experienced user who only wants to work with already existing conversion software.

On the other hand the JCAMP–DX format should not be used simply as a 'black box' but should be transparent to the user. Otherwise he or she may not really prefer the format.

For many people the difference between the 'data transfer' between computers itself and a 'data exchange format' is not quite clear and sometimes leads to confusion.

In the following chapter an attempt will be made to represent all this necessary background knowledge without boring the reader with unnecessary (although very important) details.

2 GENERAL PROBLEMS OF DATA EXCHANGE BETWEEN COMPUTER SYSTEMS

Before we go into detail, we should always keep in mind that whenever we are talking about 'data' in the following sections we always mean 'spectroscopic data' or to be even more specific 'infrared spectroscopic data'.

Our request seems to be quite simple at first glance. We want to exchange spectroscopic data between the institute *ABC* and the institute *XYZ*. Before we lived in the 'computer age' this was a really simple task. The spectrometer produced a sheet of paper with a spectrum on it. Additional information was simply hand written on this paper and the output sent by mail to our colleague. But now we have a computer connected to our spectrometer and (why not) we want to use it for the same job.

But how can we do that?

INSTITUTE ABC   INSTITUTE XYZ

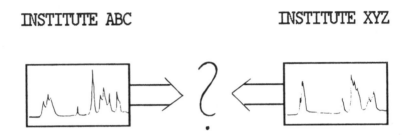

Figure 1   *How to solve this problem?*

When we want to exchange data between two or more computer systems we have to take care of two different aspects:
  (i)   Computer Hardware; The 'mechanical/electronic' coupling between computer systems.
  (ii) Computer Software; The data formats generated by application programs.

2.1 Overview about 'Mechanical/Electronic' Coupling between Computer Systems

Although it is not essential for understanding the JCAMP–DX data exchange philosophy a short overview will be given about some alternatives used to 'transport' data from one computer system to another, because nearly all of those options are limited to greater or lesser extent. The following list shows the most commonly used 'transport media'.

2.1.1   Magnetic Media.
(a) Hard Disk. A hard disk can only be used to transfer data between identical or 'compatible' computer systems. In addition it is limited to systems with removable disk cartridges, because the 'fixed' disk of a computer normally must remain where it is as part of the computer system.

(b) Floppy Diskette. Floppy diskettes are the most commonly used magnetic media for data exchange, but can only be used between identical or 'compatible' computer systems. Although the diskettes themselves and the disk drives are standardized the format in which the data are written onto the diskette can be completely different. Only for a limited number of computer systems (*e.g.* for IBM PC compatibles and APPLE MACINTOSH) do conversion programs exist which are able to read diskettes from other sources.

(c) Magnetic Tape. For magnetic tapes we find nearly the same problems as for floppy diskettes. The magnetic tapes themselves are standardized, but the data format on the tapes is not. On the other hand it is much simpler to read and convert these tapes especially on mainframe computers, so that this can really be a suitable solution.

2.1.2 Hardwired Connections. Hardwired connections are direct connections between computer systems via cables. These connections are (theoretically) standardized and in most cases easy to install. The various connections mainly differ in speed of transfer, allowed distance and price.

(a) RS-232. This is the most widely used and cheapest serial connection and is well suited for long distances. With a direct connection the cable can be up to 50 m or even longer, using a telephone modem and telephone lines the distance is unlimited and one can connect computers around the world. The average transfer rate for short distance connections is 870 bytes s^{-1} (with 9600 baud), but can rapidly decrease down to 27 bytes s^{-1} (with 300 baud) for longer distances.

One of the problems with RS-232 is that different 'standards' exist and that the computer hardware is not always reliable.

(b) IEEE-488. This is a standardized parallel connection which is well suited for fast connections over short distances. The distance is limited to about 4–5 m, but the transfer rate lies in the range of 30 Kbytes s^{-1}. This kind of connection can also be used for computer networks.

(c) ETHERNET. This (copyright by Digital Equipment) is an extremely fast serial connection for medium to long distances. The connection can be up to 2.5 km with a standard ethernet cable allowing a transfer rate of more than 1 Mbytes s^{-1}. A low cost version is also available which can be used for up to 900 m with a thinner cable. Ethernet is mainly used for large computer networks with mainframe computers.

2.2 Problems with Software and Data Formats

If we have managed to connect two computers we are immediately faced with the next and much more severe problem—different software packages for spectroscopic data manipulation or the operating software for spectrometers generate and use different data formats and only interpret their own data format, resulting in . . .

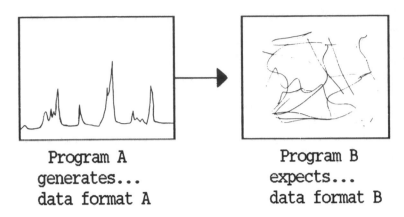

Program A
generates...
data format A

Program B
expects...
data format B

Figure 2   *Problems with data formats*

There are several solutions to get around this but the only really reasonable one is a unique format for data exchange.

To understand (and accept) this fact we must first have a look at other alternatives.

Initially the ideal solution for sure would be a single unique data format which is used worldwide in all software packages. This approach is totally unrealistic.

The only solution up to now was to supply routines for each existing software package which are able to convert data from other sources into the 'internal' format of the package. Not only the lack of support by instrument manufacturers or software houses, but also the numerous existing formats and possible changes of these formats restrict this approach to a very limited number of applications.

With a unique data exchange format each software package only needs one conversion routine which generates the internal format from the exchange format and produces the exchange format from the internal one.

3   REQUIREMENTS FOR A DATA EXCHANGE FORMAT

The verification of such a data exchange format is not so trivial as it might seem at first sight.

There are several essential and extremely important requirements for such a format which should be used and established worldwide in a variety of applications.

The data exchange format must cover the complete range of spectroscopic data, *i.e.* it should not be limited to, for example, infrared spectroscopy.

Besides the spectrum itself descriptive information about the sample, the instrument, measurement conditions, *etc.* should be part of the data set.

With respect to storage media and transfer time the format must produce a compact and small data set but this is extremely important, without reducing or changing the data itself.

The format must not only be readable for computers, but also for human beings. This also comprises the use of text programs to make additions or modifications in the data set.

The format must be designed in such a way that it can be generated and handled on all commercially available computer systems.

On these computers high level language programs must be able to manipulate the data directly—this is necessary if users want to write their own software, for example in a language like BASIC or PASCAL.

The format must state fixed rules for the programmer to inhibit any possible variation or different 'dialects' of the format.

On the other hand the format must be flexible enough and future oriented because it must also survive prospective developments.

4 REALIZATION OF THE JCAMP-DX FORMAT FOR INFRARED SPECTROSCOPY

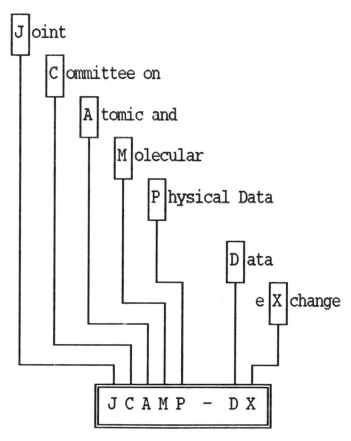

Figure 3 *What does it mean JCAMP-DX?*

The JCAMP-DX format meets all recommendations stated in the previous section.

The JCAMP-DX format is a pure ASCII text format which means it uses a character set acceptable worldwide on nearly all computer systems. Only some mainframe computers use a different character set (EBCDIC) which can be converted very easily into the ASCII format and *vice versa*.

Spectral data, peak tables, compound information, and a lot of other text information are part of the format.

The format uses intelligent algorithms to compress the spectral data without losing any information.

It is designed as an open system and can therefore easily be extended to other spectroscopic methods other than infrared spectroscopy.

The minimum contents of such a data set can be realized when we try to think about the (trivial) question:

WHAT IS SPECTROSCOPY?

—spectroscopy deals with chemical compounds, therefore we need to include:
   name or description of the compound
   chemical and physical properties
   chemical structure
   source of compound
—spectroscopy is performed with spectrometers, therefore we need to include:
   spectroscopic method
   measurement technique
   sample preparation method
   instrument type and name
   essential measurement parameters
   laboratory, institute, company where the measurement is done
—spectroscopy produces spectral data as a result, therefore we need to include:
   spectral curve
   and/or line spectrum
   and/or table of values

5   PHYSICAL STRUCTURE OF A JCAMP-DX FILE

The general structure of a JCAMP-DX data set is quite simple and logical. The description below uses and illustrates the nomenclature of the original publication.

A JCAMP-DX data set is always stored as a JCAMP-DX File (Figure 4).

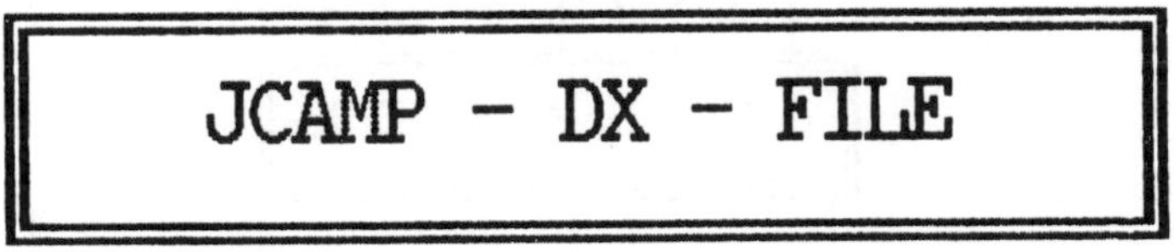

Figure 4   *JCAMP-DX File*

The JCAMP-DX File consists of a set of Labelled–Data–Record (Figure 5) which contain the data.

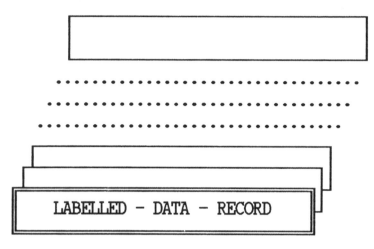

Figure 5 *A file consists of Labelled-Data-Records*

Each Labelled–Data–Record consists of a Data Label and a Data Set. The whole record is split into lines of a maximum of 80 characters, each line terminated by 'CARRIAGE RETURN' and/or 'LINEFEED' (Figure 6).

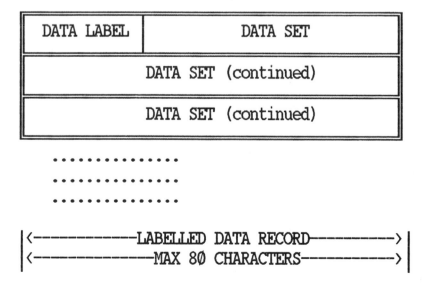

Figure 6 *Structure of a Labelled-Data-Record*

The Data Label is a text field of fixed syntax which describes the contents of the subsequent data set. The Data Label itself starts with the Data Label Flag '##' followed by the text of the Label and is terminated by the Label Terminator '='. Depending on the type of information the Data Set can either be TEXT, STRING, AFFN, or ASDF (Figure 7). The different formats of the Data Set are explained in the following section.

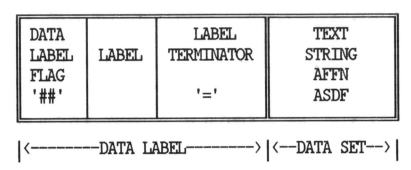

Figure 7   *Formats of data label and data set*

6   DATA FORMATS USED IN THE DATA SET

The following data formats are used within the JCAMP–DX format:
(a) TEXT. This can be any text, descriptive information, *etc.* which is not intended to be processed by computers (*e.g.* comments, *etc.*)
(b) STRING. 'STRING' is also text, but is restricted to a predefined syntax so that it can be 'understood' by computers. For example for the X units in a spectrum only a predefined set of words may be used, *e.g.* 1/CM, MICROMETERS, NANOMETERS, SECONDS.
(c) AFFN. This means 'ASCII Free Format Numeric' and can only be used for pure numeric fields. These fields must also be read by computers.
   The only allowed characters for this format are:

$$0 \quad 1 \quad 2 \quad 3 \quad 4 \quad 5 \quad 6 \quad 7 \quad 8 \quad 9 \quad + \quad - \quad . \quad , \quad E$$

(Example: -1.234 or 1.234 E 10)
   Dimensions as well as explanatory text are forbidden in these fields.
(d)ASDF. This is the acronym for 'ASCII squeezed difference form' and is a format which is exclusively used for spectral or tabular data.
   Fields with this format always start with a symbolic expression specifying the subsequent data table. The most commonly used tables are:

$(XY .. XY)$ table with pairs of X–Y values

$X^1 \, Y^1 \quad X^2 \, Y^2 \quad .... \quad Xn \, Yn$
$Xn{+}1 \, Yn{+}1 \quad .... \quad Xm \, Ym$
$......$

$(XYZ .. XYZ)$ *table with triplets of X–Y–Z values*

$$X^1\ Y^1\ Z^1\quad X^2\ Y^2\ Z^2\ .\ .\ .\ .\ .\ ,\ Xm\ Ym\ Zm$$

. . . . . .

$(X++(Y\ .\ .\ Y))$ table of Y values with equidistant $X-$
intervals

$$X^1\ Y^1\ Y^2\ .\ .\ .\ .\ Yn$$
$$Xn+1\ Yn+1\ Yn+2\ .\ .\ .\ .\ Ym$$

. . . . . .

7   LOGICAL STRUCTURE OF A JCAMP–DX FILE

The contents of a JCAMP–DX file are structured and consist of several logical blocks. The sequence of the blocks should be kept while the order inside the blocks is rather free. We distinguish between the essential part which must be present in each JCAMP–DX File and the optional part which is recommended (Figure 8).

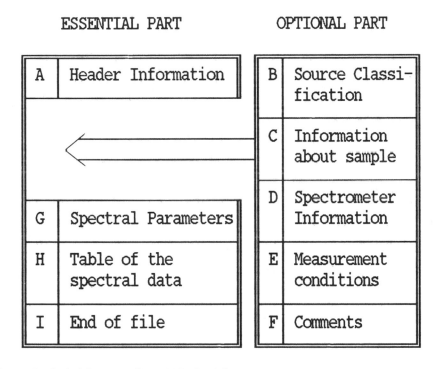

Figure 8   *Logical Structure of a JCAMP–DX File*

The following table gives an overview about the contents of the different blocks. The list is not complete, but includes the most important JCAMP–DX Labels.

A JCAMP File must always begin with the Label ##TITLE= and ends with the label ##END=.

Table 1   *Logical JCAMP–DX Blocks*

	Type of Information	Data Label (examples)
A	Header Information	##TITLE= ##JCAMP–DX= ##DATATYPE=
B	Source Classification	##ORIGIN= ##OWNER= ##DATE=
C	Information about Sample	##CAS NAME= ##MOLFORM=
D	Spectrometer Information	##SPECTROMETER/DATA SYSTEM= ##INSTRUMENT PARAMETERS=
E	Measurement Conditions	##STATE= ##TEMPERATURE=
F	Comment	##COMMENT=
G	Spectral Parameters	##XUNITS= ##YUNITS= ##FIRSTX= ##FIRSTY=
H	Table of the Spectral Data	##XYDATA= ##PEAK TABLE=
I	End of File	##END=

8   COMPRESSION OF NUMERIC DATA

One of the main advantages of the JCAMP–DX format is the compression algorithms for numeric data which fulfil the requirement of a compact format without losing any information. The compression of abscissa and ordinate values is performed in different ways.

8.1   Compression of Abscissa Values

In Fourier transform Infrared spectroscopy all spectrum points are equally spaced along the wavenumber axis in cm^{-1} with the interval derived from the frequency of the He–Ne laser. Therefore the frequency of each spectrum point is always a fractional number which can be converted to an integer by a single multiplication factor

XFACTOR=1.0	XFACTOR=1.92881
599.860	311
648.081	336
682.799	354
.	.
.	.

thus saving about 50% space.

8.2 Compression of Ordinate Values

The ordinate values are by far the largest part of a data set. The size requirement not only depends on the number of points but also on the dynamic range of the data (or word length of the computer). Therefore the ordinate compression needs to be more efficient.

By analogy with the abscissa values the ordinate values should also be converted to integers by means of a single multiplication factor for all values. The following examples therefore assume integer numbers.

For each compression format the same ten ordinate values (18, 23, 18, 15, 12, 9, 0, −5, −10, −15) are shown in a scale graph to demonstrate the efficiency of the different compression algorithms.

Five formats with different packing efficiency exist and can be used:

FIX   PAC   SQZ   DIF   DIFDUP

(a) FIX (fixed format). This is not really a compressed format because the values are printed as if they are written out by a FORTRAN program. Each ordinate value requires the same number of digits independent of its value (*e.g.* 15, 5 digits for each number).

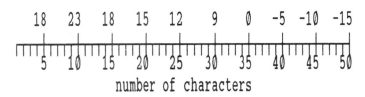

number of characters

(50 characters are needed for ten values)

Figure 9   *The FIX format*

In the author's opinion there is no reason to use this format. Even the ease of producing a data set with this format is no acceptable justification.

(b) PAC (packed format). In the PAC format all neighbouring values in a line are separated either by '+', '−', or blanks.

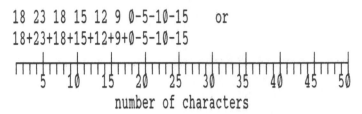

18 23 18 15 12 9 Ø-5-1Ø-15   or
18+23+18+15+12+9+Ø-5-1Ø-15

number of characters

(26 characters are needed for ten values)

Figure 10   *The PAC format*

Even this format gives no real compression and should only be used for test purposes but not as a final state.

(c) SQZ (squeezed format). The SQZ format is the first compressed format where the separators between two subsequent numbers are omitted. To recognize the beginning of a new number simply, each first digit of a number is replaced by a so called 'pseudo' number which is a letter instead of a number. So for example the number '+1' is replaced by the capital letter 'A' while '−1' is replaced by lower case 'a'. The number 1234 is then written as A234 while the number −1234 would be written as a234. The following table shows the relations between numbers and letters.

number	'pseudo' number
0 +1 +2 . . . +9	@ A B . . . I
−1 −2 . . . −9	a b . . . i

Our ten number example would read as follows:

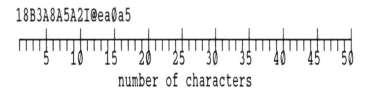

18B3A8A5A2I@eaØa5

number of characters

(17 characters are needed for ten values)

Figure 11   *The SQZ format*

As we will see later each of these formats uses a different set of 'pseudo' numbers and can thus be identified quite easily. Please note that for checking purposes the very first value of each line is always represented as a 'normal' number.

(d) DIF (difference format). The DIF format replaces the sequence of original ordinate values by the differences between two subsequent values. As mentioned above this format is identified by a different set of 'pseudo' numbers which is listed below.

number	'pseudo' number
0 +1 +2 . . . +9	% J K . . . R
−1 −2 . . . −9	j k . . . r

So instead of having four long numbers like 1234, 1237, 1254, and 1261 we use the differences 3, 17, and 7 between these numbers instead. After conversion by using 'pseudo' numbers we write these three differences as L, J7, and P. For checking purposes and prevention of errors each new line starts with a full ordinate value so that our four long numbers would now read as 1234LJ7P.

Our ten number example is now written as:

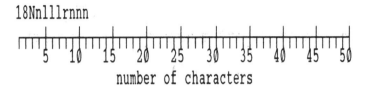

number of characters

(11 characters are needed for ten values)

Figure 12   *The DIF format*

(e) DIFDUP (difference-duplicate format). A further compression of the DIF format is achieved if repetitions of the same difference values are replaced by the value itself and a repetition count . So instead of writing five times the same number 12 12 12 12 12 we can say 12 5 which means that the number 12 occurs 5 times. The repetition count can again be distinguished from numbers by another set of 'pseudo' numbers shown below.

number	'pseudo' number
1 2 . . . 8 9	S T . . . Z s

Assuming that our number 12 is a difference we would write the above sequence as J2W.

With this most effectively compressed JCAMP–DX format our example is reduced from 50 digits for the FIX format to a total of 9 characters for the DIFDUP format.

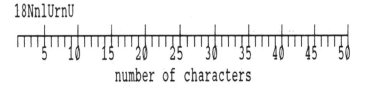

number of characters

(9 characters are needed for ten values)

Figure 13   *The DIFDUP format*

Again the very first number of a line is represented as 'normal' number and is used for error checking.

The only disadvantage of these compressed formats is the difficulty in reading these numbers. However this is not normally necessary because these tables are not intended for reading by people but for interpreting by computer programs. Nevertheless it is still possible to convert the numbers by hand if this is really desired.

9   CHEMICAL STRUCTURES IN JCAMP-DX

The original paper by McDonald and Wilks[1] does not include the coding of chemical structure formulae which is really one of the most important items of compound information. The reason for that was the criticism with the first approach which used a linear structure representation. Meanwhile a subcommittee of the 'Arbeitskreis Spektroskopie' in West Germany has developed a 'standard exchange format for chemical structure information' which is based on the JCAMP-DX format and should become part of the JCAMP-DX standard. The format is mainly based on so called 'connection tables'. More detailed information can be obtained from the authors.[2]

10   THE FUTURE OF THE JCAMP-DX FORMAT

Although the acceptance of the format is really very positive some essential points have to be observed in the future.

To keep the JCAMP-DX Format as an international standard the following activities are mandatory:

—all instrument manufacturers, companies, institutes, and departments which are working with internal formats should write or supply software for conversion and decoding.

This demand is already partially accomplished and is now dependent on the further promotion of the JCAMP-DX format.

—programmers writing conversion routines must use exactly the rules defined in the original paper.

This is not yet the case! Some programs are still based on older draft versions and have not yet been updated. Some of the JCAMP-DX labels, although described in detail in the publication, are used with a wrong format or even abused for different contents.

—programs must be easy to use to be accepted by the end user.

Much work has still be to be done. Some of the programs seem to drift towards the opposite direction—they are prohibitive for the user.

[1] R.S. McDonald and P.A. Wilks, *Appl. Spectrosc.*, 1988, **42**, 151.
[2] J. Gasteiger, B.M.P. Hendriks, P. Hoever, C. Jochum and H. Somberg 'JCAMP-CS: A Standard Exchange Format for Chemical Structure Information in Computer Readable Form', unpublished.

—programs should at least convert the data with the most compressed format (DIFDUP) to enhance the application of the format.

Most of the existing programs do not support this format but use the FIX format instead—a real drawback for the efficient use of the format in practical applications.

—the format should be adapted to other spectroscopic methods besides infrared.

This is currently under development for NMR, Mass Spectrometry, and Chromatography.

11 CONCLUSION

The JCAMP–DX Format is the first step which opens up communication in the field of spectroscopy. It will be the main criterion for the acceptance of LIMS (Laboratory Management Systems) which connect not only computers but instrumentation of different kinds and from different manufacturers. The 'off line' work with spectroscopic data on a personal computer becomes more and more important and is currently mainly limited by the complications of getting the data from the instrument to the PC.

JCAMP–DX is copyright 1986, 1987 by the Joint Committee on Atomic and Molecular Physical Data (JCAMP) and is placed in the public domain.

APPENDIX A

The JCAMP Organization

Members of the JCAMP committee are:
  ASTM
  The Coblentz Society
  The Canadian Association for Applied Spectroscopy
  The Manufacturing Chemists Association
  The American Petroleum Institute
Sponsors:
  American Chemical Society
  American Physical Society
  American Society for Mass Spectrometry
  American Society for Testing and Materials
  Coblentz Society
  Manufacturing Chemists Association
  Optical Society of America
  Society of Applied Spectroscopy
  Spectroscopy Society of Canada
Cooperating groups:
  all Instrument manufacturers
  Infrared and data specialists

Responsibility:
  Robert S. McDonald
  9 Woodside Driver
  Burnt Hills, NY 12027

APPENDIX B

Example of a JCAMP-DX File

```
##TITLE= RIFAMPICIN
##JCAMP-DX= 4.24
##DATA TYPE= INFRARED SPECTRUM
##= BRUKER ATS <--> JCAMP-DX (4.24) CONVERSION PROGRAM, VS. NW
    1.3
##MOLFORM= C43.H58.N4.O12.
##MW= 822.9500
##CAS REGISTRY NO= 13292-46-1
##SAMPLING PROCEDURE= KBR
##NAMES= 3-[[(4-METHYL-1-PIPERAZINYL)IMINO]METHYL]RIFAMYCIN
##= SPECTRUM NUMBER=NE692214
##XUNITS= 1/CM
##YUNITS= ABSORBANCE
##RESOLUTION= 1.92871
##FIRSTX= 4001.110839
##LASTX= 400.2075195
##DELTAX= -9.64355468E-1
##MAXY= 1.288641214
##MINY= 0.1745629310
##XFACTOR= 4.882812500E-4
##YFACTOR= 2.384185791E-5
##NPOINTS= 3735
##FIRSTY= 22.49352931
##XYDATA= (X++(Y..Y))
8194275I43447k13m37j80105OJ165k905Q54O62r58q92p36O57p25j29011J
382J264O24J519
```

8156750I42236L062M021J776q66j885J255N452J2641318k964j677k2m25j
1971551k688j218
8123175I39255k591r6j2561426N54J254137j901p20R25147Q82K76P21K78
J006P71547j380
8085650I29627L42J191k6j900K12K014160k189k619k661p92o82L44N85J6
43p451395L8j38
· · · · · ·
· · · · · ·

890725I19746k285n803J777Q238o630387m050J492L3517761569J0412L21
36K2978J0830O585
859125I87996J2524J4647J2192P801L1397Q5256N1526J1640L6697J20881
1944p822J0024
833450A234022j0125r416J592k6697196871713J9267
##END=

Use of the JCAMP–DX Format: A Practical Demonstration of Data Exchange

P.S. McINTYRE AND S. READE

1   INTRODUCTION

We became interested in the exchange of data between spectrometers on publication[1] of the JCAMP–DX format by the Joint Committee on Atomic and
Molecular Physical Data in 1988. The reason was at least two-fold; we had
spectrometers from three different manufacturers, and wished to create a
common database of spectra. This meant that we needed the facility of transfer
of spectra between instruments. Secondly, we wished to transmit spectra to
laboratories in other institutions, in the Netherlands and other countries, as
part of a cooperative project.

Software quickly became available from instrument manufacturers for
export of JCAMP–DX files from their spectrometers, and it was mutually
assumed that having acquired the transfer software, it would be easy to transfer spectra between instruments. This assumption turned out to be naive.

The resulting work to achieve the aims above, formed the basis of a practical
demonstration of exchange of spectra between instruments from different
manufacturers. These were the Perkin Elmer 880 dispersive spectrometer and
the Digilab FTS50 and Mattson Sirius Fourier transform spectrometers.

2   THE PROBLEM

We found very quickly that different manufacturers used different (but
allowed) JCAMP–DX formats. These formats were totally incompatible with

[1] R.S. McDonald and P.A. Wilks, *Appl. Spectrosc.*, 1988, **42**, 151.

other manufacturers software, *e.g.* a file imported from a Perkin Elmer 880 dispersive instrument would not display using Mattson Sirius software. Our Digilab FT550 instrument was also unable to display either set of JCAMP files.

We give below examples of these three formats. The spectra were of the same sample, run on the same day.

The spectrum from the Perkin Elmer instrument was run at a resolution of 3.2 cm^{-1}, whilst the others were at a resolution of 4.0 cm^{-1}. These parameters are as close as it is possible to achieve on the three instruments concerned. Also note that although the JCAMP files covered the range 400–4000 cm^{-1} (for Digilab and Mattson) and 600–4000 cm^{-1} for the Perkin Elmer instrument, we only reproduce here the data section for a band in the spectrum between 1420–1520 cm^{-1}. For an explanation of these formats see reference 2.

PERKIN ELMER: (The full file occupies 27K on disk)
##TITLE = INDENE. NEAT 0.025MM CELL.
##JCAMP-DX = 4.24
##DATA TYPE = INFRARED SPECTRUM
##ORIGIN = Perkin-Elmer Model 881 IR Spectrometer
##OWNER = P S McINTYRE and S READE
##DATE = 02-23-1990; TIME = 11:37:31
##SPECTROMETER/DATA SYSTEM = PERKIN-ELMER 880
##INSTRUMENT PARAMETERS = SLIT PROGRAM = 2 ; NOISE FILTER = 2
##DATA PROCESSING = ACCUMULATIONS = 0
##DELTAX = -1
##XUNITS = 1/CM
##YUNITS = TRANSMITTANCE
##XFACTOR = 1.00
##YFACTOR = 0.00005
##FIRSTX = 4000
##LASTX = 600
##NPOINTS = 3401
##MINY = .00575
##MAXY = .91445
##FIRSTY = .8832
##XYDATA = (X + + (Y..Y))
1520.0 + 17751 + 17744 + 17742 + 17733 + 17708 + 17678 + 17655 + 17642 + 17636
1510.0 + 17639 + 17646 + 17655 + 17661 + 17654 + 17614 + 17532 + 17422 + 17320 + 17247
1500.0 + 17198 + 17144 + 17074 + 16990 + 16870 + 16659 + 16275 + 15771 + 15301 + 15028
1490.0 + 14927 + 14914 + 14904 + 14853 + 14716 + 14517 + 14303 + 14119 + 13982 + 13907
1480.0 + 13888 + 13907 + 13933 + 13946 + 13934 + 13900 + 13865 + 13819 + 13583 + 12953
1470.0 + 11924 + 10747 + 9683 + 8900 + 8428 + 7989 + 7290 + 6221 + 5107 + 4235
1460.0 + 3743 + 3472 + 3285 + 3118 + 3127 + 3559 + 4548 + 5873 + 7213 + 8343
1450.0 + 9155 + 9646 + 10003 + 10534 + 11318 + 12216 + 13019 + 13638 + 14085 + 14409
1440.0 + 14644 + 14802 + 14912 + 15036 + 15223 + 15474 + 15711 + 15902 + 16056 + 16214
1430.0 + 16369 + 16495 + 16577 + 16628 + 16664 + 16685 + 16707 + 16716 + 16698 + 16663
1420.0 + 16651 + 16649 + 16610 + 16509 + 16374 + 16246 + 16127 + 16000 + 15853 + 15702
##END =

MATTSON: (The full file occupies 12K on disk)
##TITLE = INDENE. NEAT 0.025MM CELL.
##JCAMP_DX = 4.24
##OWNER = P S McINTYRE and S READE
##ORIGIN =
##DATA TYPE = INFRARED SPECTRUM
##YUNITS = ABSORBANCE
##XUNITS = 1/CM
##FIRSTX = 400
##FIRSTY = .161967
##LASTX = 4000
##XFACTOR = 1
##YFACTOR = .0001

```
##DELTAX = 1.92847
##NPOINTS = 1867
##XYDATA = (X + + (Y..Y))
1418.649  867  851  770  765  766  742  790  867  929  998  1167
1439.871  1233  1281  1430  1742  2283  2831  3122  3879  5276  6778  8187
1461.093  7728  6553  5319  4328  3642  2919  2373  1926  1708  1613  1566
1482.315  1565  1500  1396  1328  1395  1446  1136  916  861  776  682
1503.537  597  578  673  685  609  555  523  580  582  533  626
1524.759  564  543  546  504  508  589  575  517  630  794  755
##END =
```

DIGILAB: (The full file occupies 16K on disk)
```
##TITLE = INDENE. NEAT 0.025MM CELL.
##JCAMP-DX = 4.10
##DATA TYPE = INFRARED SPECTRUM
##ORIGIN =
##OWNER = P S McINTYRE and S READE
##DATE = 90/02/23
##TIME = 11:30:26
##CREATED = Fri Feb 23 11:30:26 1990
##SPECTROMETER/DATA SYSTEM = Digilab Data system = 3250
##XUNITS = 1/CM
##YUNITS = ABSORBANCE
##RESOLUTION = 4.0
##FIRSTX = 3.97335161e + 02
##LASTX = 4.00228378e + 03
##FIRSTY = 0.000000000e + 00
##MAXY = 7.999999996e + 00
##MINY = -6.152447686e-02
##XFACTOR = 1.92881146e + 00
##YFACTOR - 1.490116119e-08
##NPOINTS = 1870
##XYDATA = (X + + (Y..Y))
736 3283095 1240574 1128768 705274 728105 1661665 1789217 2079324 4470678
745 5861775 4102178 4102432 5281509 8130694 13267958 16154618 18214103
753 26082283 42968332 73201419 77105035 54573819 40335788 32242115 25494684
761 18952539 14648393 13066950 10353612 8290321 6996994 6413064 6620191
769 5961389 5920670 6829297 8569928 7338900 3883439 3725767 4079811 2559398
778 1056686 842317 4131530 7988674 4572662 1946585 1243907 1347526 2872174
787 2821303 4214651 4633771 2527621 2189677 1301058 264493 1361866 2625422
##END =
```

At first glance, what can we make of this?

(a) The Perkin Elmer file is in Transmittance units, the others are in Absorbance.

(b) The Perkin Elmer format has a digital resolution of 1.0 cm^{-1}, while the others have a digital resolution of 2.0 cm^{-1}. This explains why the Perkin Elmer file is about twice as large as the others.

(c) The Digilab format has compressed abscissa values, the fractional frequencies being multiplied by a constant number (##XFACTOR = 1.92881146), whilst the Mattson format uses the frequencies from the laser and therefore more space. The problem does not arise with the dispersive Perkin Elmer instrument.

(d) The ordinate values in the Perkin Elmer file are in FIX format,[2] the ##FACTOR = 00005, which gives a scale from 0 to 20000 for 0 to 100% Transmittance. These numbers are then formatted as in a FORTRAN data field,

---

[2] Exchange of Infrared data using the JCAMP–DX, H Somberg, this volume.

using 7 entries, 5 for the number, a sign and a space. This is obviously wasteful both on memory and disk space.

The ordinate values in the Mattson file are absorbance units multiplied by 10000, the ##YFACTOR = 0.00001. The numbers are 11 to a line in PAC format,[2] separated by a space, the line length in the file is therefore variable and can be as long as 85 characters, breaking the JCAMP convention of a maximum of 80 characters.

The ordinate values in the Digilab file are also in absorbance units multiplied by 6.710886402×10^7, the ##YFACTOR = $1.490116119 \times 10^{-8}$. These numbers are arranged in PAC format, separated by a space or a sign and arranged to a maximum of 80 characters to a line. This leads to a variable number of ordinate values to the line.

(e) The Perkin Elmer format is most easily read but uses more disk space. However when both memory and hard disks are cheap, does this matter?

3   A SOLUTION

The solution reached was to write a conversion program which accepted these three formats and converted them into either (or both) of the others.

The program was written in TURBO BASIC, which runs compiled, and is therefore faster. This was an important consideration because of the large amount of data to be read and written and also the large amount of arithmetic to be performed in the reformatting of the data. A copy of the completed program is appended. This is adequate for the present purpose but is capable of modification or extension for further needs.

This program allowed transfer between the instruments with no loss of quality of the spectra. We could now successfully display Perkin Elmer spectra on our Digilab or Mattson instruments and all other combinations. Example spectra are shown as Figures 1 and 2. Note however that although the spectra are correct, the scale of absorbance or transmittance has not been scaled, but the program could readily be extended to achieve this.

The program also displays spectra from JCAMP files from all three instruments (without axes, but these could be easily added).

Our second aim was to be able to transmit these spectra to other remote instruments and between laboratories in this institution. This was accomplished by using our mainframe computer and networks like JANET. All manufacturers supply software to transmit from instruments to a PC. The JCAMP files could therefore be converted on a PC and relayed to the mainframe using KERMIT. These JCAMP files could then be passed to the instruments and viewed. This was also successfully accomplished.

4   ALTERNATIVE PROCEDURES

Since this work was completed, we have purchased software from Sprouse Scientific (Micro-Search) which imports spectra from our three instruments and converts the JCAMP files to a common format (a different one again!). It does

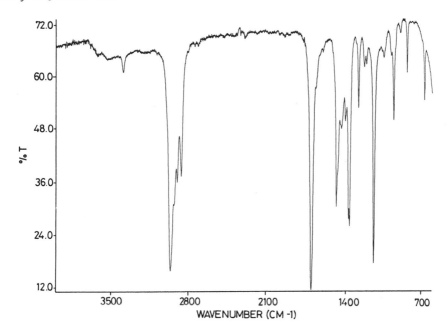

Figure 1 *Original Perkin Elmer spectrum*

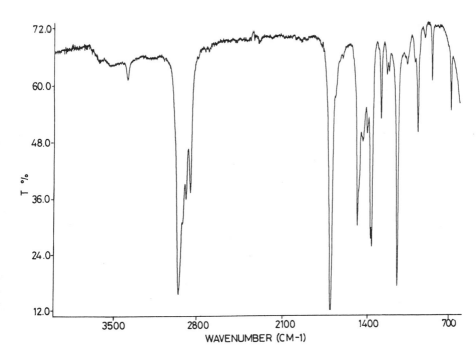

Figure 2 *Reformatted file from Digilab*

allow conversion between the formats, but does allow the building of a database from all three types of spectra (another of our aims).

We show below the same three files as above after conversion to the Sprouse format. This format combines the best points of the Mattson and Digilab formats, saves more space and does not break the 80 column rule. Again we give only a small portion of the data, roughly the same part as above.

PERKIN ELMER: (The full file now occupies 25K on disk)
##TITLE = INDENE. NEAT 0.025MM CELL.
##JCAMP-DX = 4.10
##DATATYPE = INFRARED SPECTRUM
##ORIGIN = SSS Format Spectral File
##DATE = 02/24/90
##XUNITS = 1/CM
##YUNITS = TRANSMITTANCE
##FIRSTX = 4000.000000
##LASTX = 600.000000
##FIRSTY = 0.883179
##XFACTOR = 1.0
##YFACTOR = 0.000028575897217
##NPOINTS = 3401
##XYDATA = (X + +(Y..Y))
1523.000000 31002 31043 31059 31059 31046 31043 31043 31027 30983 30930 30890
1512.000000 30867 30857 30862 30875 30890 30901 30889 30818 30675 30483 30305
1501.000000 30177 30091 29996 29874 29727 29517 29148 28475 27594 26772 26293 1490.000000 26117 26095
26077 25988 25748 25400 25026 24703 24464 24333 24300 1479.000000 24333 24378 24401 24380 24320 24259
24179 23766 22663 20863 18803 1468.000000 16941 15571 14746 13978 12754 10884 8935 7409 6548 6074 5747
5455 1456.000000 5471 6227 7957 10275 12619 14597 16018 16877 17501 18430 19802 1445.000000 21373
22779 23862 24643 25211 25622 25898 26091 26307 26635 27074 1434.000000 27488 27823 28093 28370 28640
28861 29004 29094 29157 29193 29231 1423.000000 29247 29216 29154 29133 29130 29062 28885 28649 28425
28217 27995 ##END =

MATTSON: (The full file now occupies 11K on disk)
##TITLE = INDENE - Mattson
##JCAMP-DX = 4.10
##DATATYPE = INFRARED SPECTRUM
##ORIGIN = SSS Format Spectral File
##DATE = 02/24/90
##XUNITS = 1/CM
##YUNITS = ABSORBANCE
##FIRSTX = 4000.000000
##LASTX = 400.000000
##FIRSTY = 0.022461
##XFACTOR = 1.0
##YFACTOR = 0.000054782867432
##NPOINTS = 1867
##XYDATA = (X + +(Y..Y))
1524.758842 1029 1141 972 1061 1058 953 1012 1110 1250 1227 1053 1089 1244 1499.678457 1416 1570 1671
2073 2639 2545 2423 2548 2737 2856 2857 2943 3117 1474.598071 3515 4330 5327 6648 7899 9708 11961 14105
14943 12372 9630 7080 1451.446945 5698 5167 4166 3179 2609 2337 2250 2130 1821 1695 1582 1441 1353
1426.366559 1398 1396 1404 1553 1582 1607 1725 1901 2064 2265 2629 3185 4055 ##END =

DIGILAB: (The full file now occupies 9.8K on disk)
##TITLE = INDENE.DT(1)
##JCAMP-DX = 4.10
##DATATYPE = INFRARED SPECTRUM
##ORIGIN = SSS Format Spectral File
##DATE = 02/24/90
##XUNITS = 1/CM
##YUNITS = ABSORBANCE
##FIRSTX = 4002.283780
##LASTX = 397.335161
##FIRSTY = -0.060547
##XFACTOR = 1.0

##YFACTOR = 0.000249992370605
##NPOINTS = 1870
##XYDATA = (X + +(Y..Y))
1535.333922 60 156 81 15 77 129 150 275 250 167 170 80 73 115 272 475 246 49 1500.615316 62 152 243 221
231 436 510 406 352 354 394 381 417 494 616 778 1469.754333 873 1128 1519 1920 2403 3252 4595 4362 2560
1553 1084 961 790 484 1442.750972 314 244 244 348 265 123 106 98 42 41 66 73 195 205 101 84 110 145
##END =

These files can be incorporated into a database and compared. These comparisons have recently been discussed in the scientific press.[3,4]

We give below the three spectra, as Figures 3–5, which were captured from the database as Hewlett–Packard plot files.

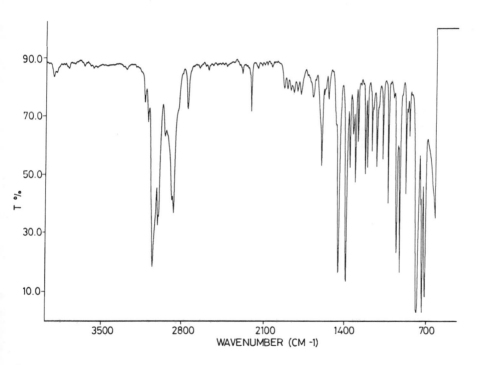

Figure 3 *Indene from Perkin Elmer JCAMP file*

[3] E.W. Ciurczak, *Spectrosc. Int.*, 1989, **1**, 12.
[4] T.W. Prouix, *Spectrosc. Int.*, 1990, **2**, 9.

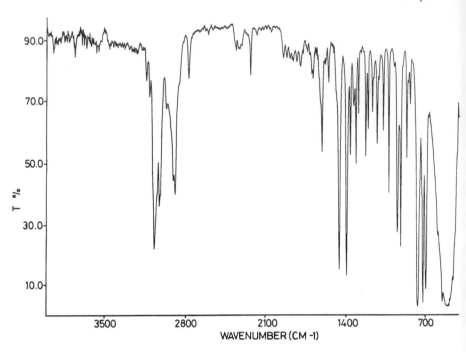

Figure 4   *Indene from Mattson JCAMP file*

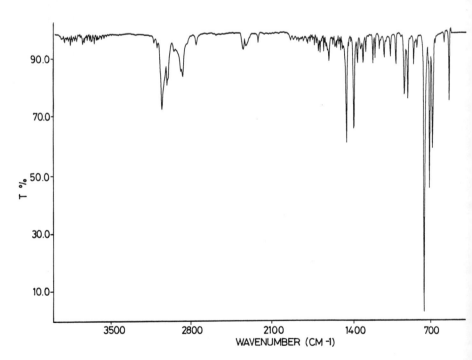

Figure 5   *Indene from Digilab JCAMP file*

5 CONCLUSION

JCAMP–DX has now been embraced by all infrared instrument manufacturers, using their own formats which make transfers between instruments of different manufacturer difficult. One manufacturer does allow import from other instruments (Bruker), which should be applauded. The solution is perhaps to purchase software that allows import from all instruments and subsequent installation of these spectra into a common database. Care is required however, in comparison of these spectra because of different instrument conditions.

APPENDIX

The File Conversion Program

```
'  LABELS%                 is the number of Labels in the Header
'  LABEL$()                contains the Header info
'  TITLE$
'  JCAMPVER$               Version of JCAMP used
'  ORIGIN$
'  OWNER$
'  DAT$                    Date (89/01/27 for Digilab), (27-01-1989 for PE)
'  TIM$                    Time
'  CREATED$                Digilab
'  INSTRUMENT$             SPECTROMETER/DATA SYSTEM
'  INSTRPARAMS$            INSTRUMENT PARAMETERS
'  DATAPROC$               DATA PROCESSING
'  DELTAX                  X value increment (applicable to PE)
'  XUNITS
'  YUNITS                  only Transmittance on PE
'  RESOLUTION              only on Digilab
'  FIRSTX
'  LASTX
'  FIRSTY
'  MAXY
'  MINY
'  LargestDiNo%            Largest Y Value in YDataVal() .... Digilab
'  LargestPeNo%            Largest Y Value in YDataVal() .... Perkin Elmer
'  LargestMa4000No%        Largest Y Value in YDataVal() .... Mattson
'  XFACTOR
'  YFACTOR
'  NoPoints%               No. of Data points
'  Mattson600%             Position in Mattson file for start of 600cm-1 range
'  Mattson4000%            Position in Mattson file for end of 4000cm-1 range
'  XYDATA$
'  XDATAVAL()              x value of the Data points
'  YDATAVAL()              raw data value
'  Num()                   first number in the row of Data
'  Filename$               name of File under inspection
'  NewFileName$            name of File to which convertion will be written
'  DI******.Dx for Digilab    PE******.Dx for Perkin-Elmer    MA******.Dx
for Mattson
'  DataLoaded

'  ***********************************************************************

On ERROR goto ERRORTRAP:
Dim LABEL$(40),Num(600),XDATAVAL(9000),YDATAVAL(9000)
Gosub Border: Gosub Title: Delay 1
Goto Options
End
```

```
Options:
        CLS
        20 Screen 8: Color 5, 3: Gosub Border
        Locate 5, 22: Print "1. List Available Spectra"
        Locate 7, 22: Print "2. Change format of Spectral JCAMP File"
        Locate 9, 22: Print "3. Display Spectra"
Locate 11, 22: Print "4. Change ROOT to spectra"
        Locate 13, 22: Print "5. EXIT"
        Locate 16, 22: Print "Please Press the Number you require "
        19 Locate 16, 59: y$=Inkey$
        If val(y$)  1 or Val(y$)  6 then Locate 16, 59: Print " ": goto 19
        If val(Y$)= 1 then Gosub Option1:
        If val(Y$)= 2 then Gosub Option2:
        If val(Y$)= 3 then Gosub Option3:
        If val(Y$)= 4 then
        Locate 18, 5: Print "Please enter new ROOT (e.g.  C: /PE881/SPECTRA/
"
                Locate 18, 52: Input Root$
                cls
        End If
        If val(Y$)= 5 then CLS: END
        If val(y$)=6 then Gosub Option6:
        Goto 20

Option1:
        Gosub Spectra:
        Gosub Returnret:
        Return

Option2:
        Close#1
        Close#2
        Gosub LoadSpectra:
        Gosub Returnret:
        If left$(Owner$, 1)="(" then Gosub Returnret: Goto 40
        Gosub Option2. 1:
        40 Return

Option2. 1:
        CLS
If Er = 53 then er=0 : goto 39
        If FileName$="" then 39
        yy$=""
        37 Screen 8: Color 5, 3: Gosub Border
        Locate 5, 18: Print "Change ......"
        if Left$(Instrument$, 6)= "PERKIN" then
                Locate 7, 28: Print "1. Perkin-Elmer to Digilab"
                Locate 8, 28: Print "2. Perkin-Elmer to Mattson"
                Locate 9, 28: Print "3. EXIT"
        elseif Left$(Instrument$, 7)= "Digilab" then
                Locate 7, 28: Print "1. Digilab to Perkin-Elmer"
                Locate 8, 28: Print "2. Digilab to Mattson"
                Locate 9, 28: Print "3. EXIT"
        elseif Left$(Instrument$, 7)= "Mattson" then
                Locate 7, 28: Print "1. Mattson to Perkin-Elmer"
                Locate 8, 28: Print "2. Mattson to Digilab"
                Locate 9, 28: Print "3. EXIT"
        end if
        Locate 13, 18: Print "Please Press the Number you require "
        38 Locate 13, 55: yy$=Inkey$
        If val(yy$)  1 or Val(yy$)  3 then Locate 13, 55: Print " ": goto 38
        if Left$(Instrument$, 6)= "PERKIN" and val(yy$) = 1 then Gosub PECASE:
        if Left$(Instrument$, 6)= "PERKIN" and val(yy$) = 2 then Gosub PECASE:
        if Left$(Instrument$, 7)= "Digilab" and val(yy$) = 1 then Gosub DICASE:
```

```
           if Left$(Instrument$, 7)= "Digilab" and val(yy$) = 2 then Gosub DICASE:
           if Left$(Instrument$, 7)= "Mattson" and val(yy$) = 1 then Gosub MACASE:
           if Left$(Instrument$, 7)= "Mattson" and val(yy$) = 2 then Gosub MACASE:
           if val(yy$) = 3 then cls
           39 Return


Option3:
           Close#1
           Close#2
           yy$="Option3"
           Gosub LoadSpectra:
           If FileName$="" then 54
           Gosub DrawSpectra:
           54 Return

Option6:
           Cls
           80 Files Root$+"*. DXR"
           81 Files Root$+"*. DXP"
           Gosub Returnret:
           Return

End

'   *******************************************************************

ClearVariables:
           Close#1
           Close#2
           Title$=""
           JCAMPVer$=""
           DataType$=""
           Origin$=""
           Owner$=""
           Created$=""
           Instrument$=""
           DataProc$=""
           DeltaX=0
           XUnits$=""
           YUnits=0
           Resolution=0
           FirstX=0
           LastX=0
           FirstY=0
           MaxY=0
           MinY=0
           XFactor=0
           YFactor=0
           NoPoints%=0
           NewFileName$=""
           Return

Spectra:
           Screen 0
           Color 2, 0
           Locate 2, 3: Print "The following is list of Spectra in JCAMP format
that are either available for"
           Locate 4, 3: Print "conversion from one format to the  other or display. "
           Locate 8, 5: print "";: Files ROOT$+"*. DX"
           Return

LoadSpectra:
           Gosub Spectra:
```

```
        Color 3, 0
        5 Locate 19, 58: Print "                             "
        Locate 19, 5: Print "Please enter<<Filename>> of Spectra (e. g.  JPOLYSS
) ";: Input   Filename$
        If   Filename$="" then 21
        If Right$(FileName$, 3) = ". DX" or Right$(FileName$, 3) = ". dx" then
                goto 31
        elseif Right$(FileName$, 4)=". DXR" or Right$(FileName$, 4)= ". dxr"
or Right$(FileName$, 4)= ". DXP" or Right$(FileName$, 4)= ". dxp" then
                goto 31
        End If
        FileName$=FileName$+". DX"
        31 Gosub ClearVariables:
        FileName$=ROOT$+Filename$
        Gosub FileInfo:
        Gosub Displaydetails:
        21 Return

FileInfo:
        Color 10, 0
        Infoin$="Please wait I am selecting information on the file "+Filename$
        If Len(Infoin$)<<=78 then
                Locate 22, (80-len(Infoin$))/2: Print Infoin$
        elseif Len(Infoin$) = 79 then
                Locate 22, 1: Print Infoin$
        End If
        open  Filename$ for input as #1
        I1%=1
        4 LINE INPUT#1, LABEL$(I1%)
        if left$(LABEL$(i1%), 8) = "##XYDATA" then 1
        I1% = I1%+1
        Goto 4
        1 close#1
        Labels% = I1%
        Gosub FileConstants:
        If Title$="" then Title$=DataType$
        open  Filename$ for input as #1
        for I1%=1 to Labels%
                LINE INPUT#1, Temp$
        next I1%
        NoOfDataLines% = 1
        2 If EOF(1) then 3


                LINE INPUT#1, Temp$
                NoOfDataLines% = NoOfDataLines% + 1
        goto 2
        3 close#1
        Return

FileConstants:
        For I1%=1 to Labels%
                if left$(Label$(I1%), 9)  =
"##TITLE= "    then Title$ = mid$(Label$(I1%), 10, (len(Label$(i1%))-9))
                if left$(Label$(I1%), 12) =
"##JCAMP-DX= " then JCAMPVER$ = mid$(Label$(I1%), 13, (len(Label$(i1%))-12))
                if left$(Label$(I1%), 12) =
"##JCAMP_DX= " then JCAMPVER$ = mid$(Label$(I1%), 13, (len(Label$(i1%))-12))
                if left$(Label$(I1%), 13) =
"##DATA TYPE= " then DATATYPE$ = mid$(Label$(I1%), 14, (len(Label$(i1%))-13))
                if left$(Label$(I1%), 10) =
"##ORIGIN= "  then ORIGIN$ = mid$(Label$(I1%), 11, (len(Label$(i1%))-10))
                if left$(Label$(I1%), 9)  =
"##OWNER= "   then OWNER$ = mid$(Label$(I1%), 10, (len(Label$(i1%))-9))
                if left$(Label$(I1%), 11) =
```

```
"##CREATED= " then CREATED$ = mid$(Label$(I1%),12,(len(Label$(i1%))-11))
              if left$(Label$(I1%),28) =
"##SPECTROMETER/DATA SYSTEM= " then INSTRUMENT$
= mid$(Label$(I1%),29,(len(Label$(i1%))-28))
              if left$(Label$(I1%),17) =
"##DATA PROCESSING= " then DATAPROC$
= mid$(Label$(I1%),18,(len(Label$(i1%))-17))
              if left$(Label$(I1%),10) =
"##DELTAX= "  then DELTAX = val(mid$(Label$(I1%),11,7))
              if left$(Label$(I1%),10) =
"##XUNITS= "  then XUNITS$ = mid$(Label$(I1%),11,(len(Label$(i1%))-10))
              if left$(Label$(I1%),10) =
"##YUNITS= "  then YUNITS$ = mid$(Label$(I1%),11,(len(Label$(i1%))-10))
              if left$(Label$(I1%),14) =
"##RESOLUTION= " then RESOLUTION = val(mid$(Label$(I1%),15,7))
              if left$(Label$(I1%),9)  =
"##FIRSTX="   then FIRSTX = val(mid$(Label$(I1%),10,17))
              if left$(Label$(I1%),8)  =
"##LASTX="    then LASTX = val(mid$(Label$(I1%),9,17))
              if left$(Label$(I1%),9)  =
"##FIRSTY="   then FIRSTY = val(mid$(Label$(I1%),10,17))
              if left$(Label$(I1%),7)  =
"##MAXY="     then MAXY = val(mid$(Label$(I1%),8,17))
              if left$(Label$(I1%),7)  =
"##MINY="     then MINY = val(mid$(Label$(I1%),8,17))
              if left$(Label$(I1%),10) =
"##XFACTOR="  then XFACTOR = val(mid$(Label$(I1%),11,17))
              if left$(Label$(I1%),10) =
"##YFACTOR="  then YFACTOR = val(mid$(Label$(I1%),11,17))
              if left$(Label$(I1%),10) =
"##NPOINTS="  then NOPOINTS% = val(mid$(Label$(I1%),11,6))
              if left$(Label$(I1%),10) =
"##XYDATA= "  then XYDATA$ = mid$(Label$(I1%),11,(len(Label$(i1%))-10))
              if Left$(Instrument$,6)= "PERKIN" then Gosub PerkinDatTim:
              if Left$(Instrument$,7)= "Digilab" then Gosub DigilabDatTim:
         Next I1%
         If Left$(Instrument$,6)=
"PERKIN" then Instrument$=Left$(Instrument$,Len(Instrument$)-1)+"1"
         If Left$(Title$,8)=" MATTSON" then Instrument$="Mattson"
         Return

DigilabDatTim:
         Dat$=Mid$(Label$(6),9,10)
         Tim$=Mid$(Label$(7),9,10)
         return

PerkinDatTim:
         Dat$=Mid$(Label$(6),8,11)
         Tim$=Mid$(Label$(6),27,9)
         Created$=" "+Tim$+" "+Dat$
         return

Displaydetails:
         color 4,2
         cls
         If Y$="3" then 41
         If Left$(Owner$,1)="(" then
              Locate 7,25: Print "BEWARE THIS FILE HAS BEEN CONVERTED"
              Locate 12,14: print Mid$(Owner$,4,Len(Owner$))
              Locate 14,13: Print "The original
filename would most likely begin
with ";Mid$(Filename$,Len(Root$)+3,6)
Locate 16,6: Print "At this stage you will not be allowed to change
the format of this File"
```

```
                Gosub Returnret
                Color 4, 2
                cls
        End If
        41 Locate 2, ((80-len(Title$))/2): print Title$
        Locate 4, 5: Print "This "; DATATYPE$;
if Left$(Instruments$, 7)="Digilab" then Print " created on "; Created$;"
using a"
if Left$(Instruments$, 7)="Mattson" then Print "was created using a"
if Left$(Instruments$, 6)="PERKIN" then Print " is dated "; Dat$;" and
was created using a"
        Locate 6, 5: Print Instruments$;
Print ", it contains information between ";
Print using "#####. ##"; FIRSTX;
Print "cm-1"
        Locate 8, 5: Print "and ";: Print using "#####. ##"; LASTX;
Print "cm-1"
        Locate 10, 5: Print "It consists of"; NoPoints%; "data points"
        Return


DICASE:
        cls
        Gosub Border:
        Locate 3, ((80-len(Title$))/2): print Title$
        LineNo1%=5: LineNo2%=6
        Gosub DigilabDataLoad:
        Gosub FileNameOptions:
        Locate 14, 5: Print "Do you want the file to consist of Raw or % data"
        Locate 15, 5: Print "Please Press either R or P "
        50 RaworPercent$=Inkey$
        If RaworPercent$="R" or RaworPercent$="r" or RaworPercent$="P" or
RaworPercent$="p" then 49 else 50
        49 If RaworPercent$="R" or RaworPercent$=
"r" then NewFileName$=NewFileName$+"R"
        If RaworPercent$="P" or RaworPercent$=
"p" then NewFileName$=NewFileName$+"P"
        Owner$ ="(D) This file has been converted from the Digilab format"
        If val(yy$)=1 then Gosub PerkinOutPut:
        If val(yy$)=2 then Gosub MattsonOutPut:
        Return


PECASE:
        cls
        Gosub Border:
        Locate 3, ((80-len(Title$))/2): print Title$
        LineNo1%=5: LineNo2%=6
        Gosub PerkinDataLoad:
        Gosub FileNameOptions:
        Locate 14, 5: Print "Do you want the file to consist of Raw or % data"
        Locate 15, 5: Print "Please Press either R or P "
        46 RaworPercent$=Inkey$
        If RaworPercent$="R" or RaworPercent$="r" or RaworPercent$="P" or
RaworPercent$="p" then 45 else 46
        45 If RaworPercent$="R" or RaworPercent$=
"r" then NewFileName$=NewFileName$+"R"
        If RaworPercent$="P" or RaworPercent$=
"p" then NewFileName$=NewFileName$+"P"
        Owner$ ="(P) This file has been converted from the Perkin-Elmer format"
        If val(yy$)=1 then Gosub DigilabOutPut:
        If val(yy$)=2 then Gosub MattsonOutPut:
        Return


MACASE:
        cls
```

```
        Gosub Border
        Locate 3, ((80-len(Title$))/2): print Title$
        LineNo1%=5: LineNo2%=6
        Gosub MattsonDataLoad:
        Gosub FileNameOptions:
        Locate 14, 5: Print "Do you want the file to consist of Raw or % data"
        Locate 15, 5: Print "Please Press either R or P "
        48 RaworPercent$=Inkey$
        If RaworPercent$="R" or RaworPercent$="r" or RaworPercent$="P" or
RaworPercent$="p" then 47 else 48
        47 If RaworPercent$="R" or RaworPercent$=
"r" then NewFileName$=NewFileName$+"R"
        If RaworPercent$="P" or RaworPercent$=
"p" then NewFileName$=NewFileName$+"P"
        Owner$ ="(M) This file has been converted from the Mattson format"
        If val(yy$)=1 then Gosub PerkinOutPut:
        If val(yy$)=2 then Gosub DigilabOutPut:
        Return


DigilabDataLoad:
        SmallestDiNo=1E+10: LargestDiNo=1E-10
        If yy$="Option3" then LineNo1%=15: LineNo2%=16
                Locate LineNo1%, 5: Print
"Please wait I am loading in your data points
"
                Locate LineNo2%, 5: Print "Number completed =        =        %"
        open  Filename$ for input as #1
        for I1% = 1 to Labels%
                LINE INPUT#1, Temp$
Next I1%
I5%=1
for I1% = 1 to (NoOfDataLines%-2)
                Tmp$=""
        Tmp1$=""
                I3%=1
LINE INPUT#1, Temp$
                for I2% = 1 to len(Temp$)
                Tmp$=mid$(Temp$, I2%, 1)
        if Tmp$=" " or Tmp$="-" or Tmp$=
"+" then TempNo(I3%)=val(Tmp1$): I3%=I3%+1: Tmp1$=""
Tmp1$ = Tmp1$+Tmp$
next I2%
                TempNo(I3%)=val(Tmp1$)
                Num(I1%)=TempNo(1)
                for I4%=2 to I3%
                        XDATAVAL(i5%)=FIRSTX+((I5%-1)*XFACTOR)
                        if i5%=1 then XDATAVAL(1)=FirstX
YDATAVAL(i5%) = TempNo(I4%)
                        If LargestDiNo  YDataVal(I5%)
then LargestDiNo= YDataVal(I5%)
                        If SmallestDiNo  YDataVal(I5%)
then SmallestDiNo=YDataVal(I5%)
                        I5%=I5%+1
                        If yy$="Option3" then LineNo1%=15: LineNo2%=16
                                Locate LineNo2%, 23: print I5%-1
                                Locate LineNo2%, 32: print
using "###. #"; ((I5%-1)/NoPoints%)*100
                next I4%
        next I1%
        return


PerkinDataLoad:
        SmallestPeNo=1E+10: LargestPeNo=1E-10
        If yy$="Option3" then LineNo1%=15: LineNo2%=16
```

```
                    Locate LineNo1%, 5: Print
"Please wait I am loading in your data points
"
                    Locate LineNo2%, 5: Print "Number completed =           =          %"
        open  Filename$ for input as #1
        for I1% = 1 to Labels%
                    LINE INPUT#1, Temp$
Next I1%
I5%=1
for I1% = 1 to (NoOfDataLines%-2)
                    Tmp$=""
        Tmp1$=""
                    I3%=1
LINE INPUT#1, Temp$
                    for I2% = 1 to len(Temp$)
                    Tmp$=mid$(Temp$, I2%, 1)
        if Tmp$="-" or Tmp$="+" then TempNo(I3%)=val(Tmp1$): I3%=I3%+1: Tmp1$=""
Tmp1$ = Tmp1$+Tmp$
next I2%
                    TempNo(I3%)=val(Tmp1$)
                    Num(I1%)=TempNo(1)
                    for I4%=2 to I3%
                           XDATAVAL(i5%)=FIRSTX+((I5%-1)*DELTAX)
                           if i5%=1 then XDATAVAL(1)=FirstX
YDATAVAL(i5%) = TempNo(I4%)
                           If LargestPENo  YDataVal(I5%)
then LargestPENo= YDataVal(I5%)
                           If SmallestPeNo  YDataVal(I5%)
then SmallestPeNo=YDataVal(I5%)
                           I5%=I5%+1
                           If yy$="Option3" then LineNo1%=15: LineNo2%=16
                                  Locate LineNo2%, 23: print I5%-1
                                  Locate LineNo2%, 32: print
using "###. #";((I5%-1)/NoPoints%)*100
                    next I4%
        next I1%
        return


MattsonDataLoad:
        SmallestMa600No=1E+10: LargestMa4000No=1E-10
        SmallestMa400No=1E+10: LargestMa64000No=1E-10
        If yy$="Option3" then LineNo1%=15: LineNo2%=16
                    Locate LineNo1%, 5: Print
"Please wait I am loading in your data points "
                    Locate LineNo2%, 5: Print "Number completed =           =          %"
        open  Filename$ for input as #1
        for I1% = 1 to Labels%
                    LINE INPUT#1, Temp$
Next I1%
I5%=1
for I1% = 1 to (NoOfDataLines%-2)
                    Tmp$=""
        Tmp1$=""
                    I3%=1
LINE INPUT#1, Temp$
                    If Left$(Temp$, 6)="##END=" then I%=NoDataLines%
                    for I2% = 1 to len(Temp$)
                    Tmp$=mid$(Temp$, I2%, 1)
        if Tmp$=" " then Tmp%=Tmp%+1
        if Tmp%=2 then TempNo(I3%)=val(Tmp1$): I3%=I3%+1: Tmp1$="": Tmp%=0
Tmp1$ = Tmp1$+Tmp$
next I2%
                    Num(I1%)=TempNo(1)
                    for I4%=2 to I3% - 1
```

```
                          XDATAVAL(i5%)=FIRSTX+((I5%-1)*DeltaX)
                          If XDataVal(I5%) <M=> 402 then Mattson400%=I5%
                          If XDataVal(I5%) <M=> 602 then Mattson600%=I5%
                          If XDataVal(I5%) <M=> 4000 then Mattson4000%=I5%
                          if i5%=1 then XDATAVAL(1)=FirstX
                          YDATAVAL(i5%) = TempNo(I4%)
                          If XDataVal(I5%) = 602 and XDataVal(I5%)   4000
then
                                  If LargestMa4000No   YDataVal(I5%)
then LargestMa4000No
= YDataVal(I5%)
                                  If SmallestMa600No   YDataVal(I5%)
then SmallestMa600No = YDataVal(I5%)
                          End If
                          If XDataVal(I5%) = 402 and XDataVal(I5%)   4000
then
                                  If LargestMa64000No   YDataVal(I5%)
then LargestMa64000No
= YDataVal(I5%)
                                  If SmallestMa400No   YDataVal(I5%)
then SmallestMa400No = YDataVal(I5%)
                          End If
                          I5%=I5%+1
                          If yy$="Option3" then LineNo1%=15: LineNo2%=16
                                  Locate LineNo2%, 23: print I5%-1
                                  Locate LineNo2%, 32: print
using "###.#";((I5%-1)/NoPoints%)*100
                  next I4%
          next I1%
Return

FileNameoptions:
          Temp1$=Mid$(Filename$,Len(ROOT$)+1,Len(Filename$)-Len(Root$)-3)
          if Left$(Instrument$,6)= "PERKIN"
and val(yy$) = 1 then NewFileName$="DI"+Left$(Temp1$,6)+".DX"
          if Left$(Instrument$,6)= "PERKIN"
and val(yy$) = 2 then NewFileName$="MA"+Left$(Temp1$,6)+".DX"
          if Left$(Instrument$,7)= "Digilab" and val(yy$) = 1
then NewFileName$="PE"+Left$(Temp1$,6)+".DX"
          if Left$(Instrument$,7)= "Digilab" and val(yy$) = 2
then NewFileName$="MA"+Left$(Temp1$,6)+".DX"
          if Left$(Instrument$,7)= "Mattson" and val(yy$) = 1
then NewFileName$="PE"+Left$(Temp1$,6)+".DX"
          if Left$(Instrument$,7)= "Mattson" and val(yy$) = 2
then NewFileName$="DI"+Left$(Temp1$,6)+".DX"
          Locate 8,5: Print "The default for the new file name will be -----
"; NewFileName$;" "
          Locate 9,5: Print "Is this new name acceptable (Enter Y or N) ";
          15 x$=Inkey$
          If x$="Y" or x$="y" then 13
          If x$="N" or x$="n" then 14 else 15
          14 Locate 11,5: Input "Please enter a 6 character string ";Temp2$
          if Left$(Instrument$,6)= "PERKIN" and val(yy$) = 1
then NewFileName$="DI"+Left$(Temp2$,6)+".DX"
          if Left$(Instrument$,6)= "PERKIN" and val(yy$) = 2
then NewFileName$="MA"+Left$(Temp2$,6)+".DX"
          if Left$(Instrument$,7)= "Digilab" and val(yy$) = 1
then NewFileName$="PE"+Left$(Temp2$,6)+".DX"
          if Left$(Instrument$,7)= "Digilab" and val(yy$) = 2
then NewFileName$="MA"+Left$(Temp2$,6)+".DX"
          if Left$(Instrument$,7)= "Mattson" and val(yy$) = 1
then NewFileName$="PE"+Left$(Temp2$,6)+".DX"
          if Left$(Instrument$,7)= "Mattson" and val(yy$) = 2
then NewFileName$="DI"+Left$(Temp2$,6)+".DX"
```

```
        Locate 12,5: Print "The New File will now be
called ----- "; NewFileName$;"
"

        NewFileName$=ROOT$+NewFileName$
        13 return

PerkinOutPut:
        open NewFileName$ for output as #2
        Print#2,"##TITLE= ";Title$
        Print#2,"##JCAMP-DX= ";JCAMPVER$
        Print#2,"##DATA TYPE= ";DATATYPE$
        Print#2,"##ORIGIN= ";ORIGIN$
        Print#2,"##OWNER= ";Owner$
        Print#2,"##DATE= ";DATE$;";  TIME= ";TIME$
        Print#2,"##SPECTROMETER/DATA SYSTEM= ";"PERKIN-ELMER" 'INSTRUMENT$
        Print#2,"##INSTRUMENT PARAMETERS= SLIT PROGRAM=  1 ;  NOISE FILTER=  2"
        Print#2,"##DATA PROCESSING= ACCUMULATIONS=  1"
        Print#2,"##DELTAX= ";:Print#2 ,using "##.##";XDataVal(1)-XDataVal(2)
        Print#2,"##XUNITS= ";XUNITS$
        Print#2,"##YUNITS= ";YUNITS$
        Print#2,"##XFACTOR= 1.00"
        If Left$(Instrument$,7)="Digilab" then Print#2,"##YFACTOR= 0.00005"
        If Left$(Instrument$,7)="Mattson"
then Print#2,"##YFACTOR= 0.00005"  'Print#2
,using "##.#####";YFACTOR
        If Left$(Instrument$,7)="Digilab" then Print#2,"##FIRSTX= ";:Print#2
,using "####";LASTX
        If Left$(Instrument$,7)="Mattson"  then Print#2,"##FIRSTX= ";:Print#2
,using "####.#";XDataVal(Mattson4000%)
        If Left$(Instrument$,7)="Digilab" then Print#2,"##LASTX= ";:Print#2
,using "####";FIRSTX
        If Left$(Instrument$,7)="Mattson"  then Print#2,"##LASTX= ";:Print#2
,using "####.#";XDataVal(Mattson600%)
        If Left$(Instrument$,7)="Digilab" then Print#2,"##NPOINTS= ";NoPoints%
        If Left$(Instrument$,7)="Mattson"
then Print#2,"##NPOINTS= ";Mattson4000%-Mattson600%+1
        If Left$(Instrument$,7)="Digilab" then Print#2,"##MINY= ";:Print#2
,using "##.#####";MINY
        If Left$(Instrument$,7)="Mattson"  then Print#2,"##MINY= ";:Print#2
,using "##.#####";SmallestMa600No*YFACTOR
        If Left$(Instrument$,7)="Digilab" then Print#2,"##MAXY= ";:Print#2
,using "##.#####";MaxY
        If Left$(Instrument$,7)="Mattson"  then Print#2,"##MAXY= ";:Print#2
,using "##.#####";LargestMa4000No*YFactor

        If Left$(Instrument$,7)="Digilab" then Print#2,"##FIRSTY= ";:Print#2
,using "##.#####";FIRSTY
        If Left$(Instrument$,7)="Mattson"  then Print#2,"##FIRSTY= ";:Print#2
,using "##.#####";YDataVal(Mattson4000%)*YFACTOR
        Print#2,"##XYDATA= ";XYDATA$
        Temp$=""
        Temp1$=""
        Temp2$=""
        Temp3$=""
        Locate 17,5: Print "Please wait I am
writing your data to file ";NewFileName$
        Locate 18,5: Print "Number completed =        =       %"
        If Left$(Instrument$,7)="Digilab" then
                I1%=NoPoints%
                34 Print#2,using "####.#";XDataVal(I1%);
                for i2%= 1 to 10
                        Temp1$=Str$(int((YDATAVAL(I1%)*YFactor)/0.00005))
                        If left$(Temp1$,1) = " "
then Temp1$=Right$(Temp1$,Len(Temp1$)-1)
else Temp1$="       "+Temp1$
```

```
                        If left$(Temp1$, 1)   "-" then
Temp1$="          +"+Temp1$ else Temp1$="          "+Temp1$
                        Temp1$=Right$(Temp1$, 7)
                        Temp2$=Temp2$+Temp1$
                        if i1%=1 then i2%=10
                        Locate 18, 23: print NoPoints%-I1%+1
                        Locate 18, 32: print
using "###. #";((NoPoints%-I1%+1)/NoPoints%)*100
                        I1% = I1% - 1
                next I2%
                Print#2, Temp2$
                Temp$=""
                Temp1$=""
                Temp2$=""
                Temp3$=""
                if i1%=0 then 35 else  34
                35 Print#2,"##END="
                Close#2
        ElseIf Left$(Instrument$, 7)="Mattson" then
                I1%=Mattson4000%
                66 Print#2, using "####. #";XDataVal(I1%);
                for i2%= 1 to 10
                        Temp1$=Str$(int((YDATAVAL(I1%)*YFactor)/0. 00005))
                        If left$(Temp1$, 1)   "-"
then Temp1$="          +"+mid$(Temp1$, 2, len(Temp1$)-1)
else Temp1$="          "+Temp1$
                        Temp1$=Right$(Temp1$, 7)
                        Temp2$=Temp2$+Temp1$
                        if i1%<M=>attson600% then i2%=10
                        Locate 18, 23: print Mattson4000%-I1%+1
                        Locate 18, 32: print
using "###. #";((Mattson4000%-I1%+1)/(Mattson4000%-Mattson600%))*100
                        I1% = I1% - 1
                next I2%
                Print#2, Temp2$

                Temp$=""
                Temp1$=""
                Temp2$=""
                Temp3$=""
                if i1%  <M=>attson600% then 65 else   66
                65 Print#2,"##END="
                Close#2
        End If
        Gosub Returnret:
        Return

DigilabOutPut:
        open  NewFileName$ for output as #2
        Print#2,"##TITLE= ";Title$
        Print#2,"##JCAMP-DX= ";JCAMPVER$
        Print#2,"##DATA TYPE= ";DATATYPE$
        Print#2,"##ORIGIN= ";ORIGIN$
        Print#2,"##OWNER= ";Owner$
        Print#2,"##DATE= ";DATE$
        Print#2,"##TIME= ";TIME$
        Print#2,"##CREATED= ";CREATED$
        Print#2,"##SPECTROMETER/DATA SYSTEM= ";"Digilab" 'INSTRUMENT$
        Print#2,"##XUNITS= ";XUNITS$
        If Left$(Instrument$, 6)="PERKIN" then Print#2,"##YUNITS= ";YUNITS$
        If Left$(Instrument$, 7)="Mattson" then
                If YUNITS$="INTENSITY"
then Print#2,"##YUNITS= TRANSMITTANCE" else
Print#2,"##YUNITS= ABSORBANCE"
        End If
```

```
        Print#2,"##RESOLUTION= x.x" ' RESOLUTION
        If Left$(Instrument$,6)="PERKIN" then Print#2,"##FIRSTX= ";:Print#2
,using "#.########^^^^";LASTX
        If Left$(Instrument$,7)="Mattson" then Print#2,"##FIRSTX= ";:Print#2
,using "#.########^^^^";FirstX ' XDataVal(Mattson400%)
        If Left$(Instrument$,6)="PERKIN" then Print#2,"##LASTX= ";:Print#2
,using "#.########^^^^";FIRSTX
        If Left$(Instrument$,7)="Mattson" then Print#2,"##LASTX= ";:Print#2
,using "#.########^^^^";LastX ' XDataVal(Mattson4000%)
        If Left$(Instrument$,6)="PERKIN" then Print#2,"##FIRSTY= ";:Print#2
,using "#.#########^^^^";YDATAVAL(NoPoints%)*YFACTOR
        If Left$(Instrument$,7)="Mattson" then Print#2,"##FIRSTY= ";:Print#2
,using "#.#########^^^^";FirstY ' YDATAVAL(Mattson600%)*YFACTOR
        If Left$(Instrument$,6)="PERKIN" then Print#2,"##MAXY= ";:Print#2
,using "#.#########^^^^";MAXY
        If Left$(Instrument$,7)="Mattson" then Print#2,"##MAXY= ";:Print#2
,using "#.#########^^^^";LargestMa64000No*YFactor
        If Left$(Instrument$,6)="PERKIN" then Print#2,"##MINY= ";:Print#2
,using "#.#########^^^^";MINY
        If Left$(Instrument$,7)="Mattson" then Print#2,"##MINY= ";:Print#2
,using "#.#########^^^^";SmallestMa400No*YFactor
        If Left$(Instrument$,6)="PERKIN" then Print#2,"##XFACTOR= ";:Print#2
,using "#.########^^^^";XFACTOR*Abs(DeltaX)
        If Left$(Instrument$,7)="Mattson" then Print#2,"##XFACTOR= 1.00000000E+00"
' Print#2 ,using "#.########^^^^";XFACTOR
        Print#2,"##YFACTOR=";:Print#2 ,using "#.########^^^^";YFACTOR
        If Left$(Instrument$,6)="PERKIN" then Print#2,"##NPOINTS=";NOPOINTS%
        If Left$(Instrument$,7)="Mattson"
then Print#2,"##NPOINTS=";Mattson4000%-Mattson400%+1
        Print#2,"##XYDATA= ";XYDATA$
        Temp$=""
        Temp1$=""
        Temp2$=""
        Temp3$=""
        Locate 17,5: Print "Please wait I am writing
your data to file ";NewFileName$
        Locate 18,5: Print "Number completed =        =       %"
        If Left$(Instrument$,6)="PERKIN" then
                I1%=NoPoints%
                16 Temp$=Str$(INT(XDATAVAL(I1%)/Abs(DeltaX)))
                Temp$=Mid$(Temp$,2,Len(Temp$))
                for i2%= 1 to 8
                        Temp1$=Str$(int(YDATAVAL(I1%)))
                        Temp2$=Temp2$+Temp1$
                        if i1%=1 then i2%=8
                        Locate 18,23: print NoPoints%-I1%+1
                        Locate 18,32: print
using "###.#";((NoPoints%-I1%+1)/NoPoints%)*100
                        I1% = I1% - 1
                next I2%
                Temp3$=Temp3$+Temp2$
                Print#2, Temp3$
                Temp$=""
                Temp1$=""
                Temp2$=""
                Temp3$=""
                if i1%=0 then 17 else  16
                17 Print#2,"##END="
                Close#2
        ElseIf Left$(Instrument$,7)="Mattson" then
                I1%=1 ' Mattson400%
                71 Temp$=mid$(Str$(XDATAVAL(I1%)),1,7)
                Temp$=Mid$(Temp$,2,Len(Temp$))
                for i2%= 1 to 10
```

```
                         Temp1$=Str$(int(YDATAVAL(I1%)))
                         Temp2$=Temp2$+Temp1$
                         if i1%=NoPoints% then I2%=10 ' Mattson4000% then i2%=10
                         Locate 18,23: print I1%
                         Locate 18,32: print
using "###. #"; ((I1%)/(NoPoints%))*100
                         I1% = I1% + 1
                next I2%
                Temp3$=Temp$+Temp2$
                Print#2, Temp3$
                Temp$=""
                Temp1$=""
                Temp2$=""
                Temp3$=""
                if i1%  NoPoints% then 72 else 71 ' Mattson4000%
then 72 else  71
                72 Print#2, "##END="
                Close#2
        End If
        Gosub Returnret:
        Return

MattsonOutPut:
        open NewFileName$ for output as #2
        Print#2, "##TITLE= "; Title$
        Print#2, "##JCAMP_DX= "; JCAMPVER$
        Print#2, "##OWNER= "; OWNER$
        Print#2, "##ORIGIN= "; ORIGIN$
        Print#2, "##SPECTROMETER/DATA SYSTEM= "; "Mattson" ' INSTRUMENT$
        Print#2, "##DATA TYPE= "; DATATYPE$
        Print#2, "##YUNITS= "; YUNITS$
        Print#2, "##XUNITS= "; XUNITS$
        If Left$(Instrument$, 7)="Digilab" then Print#2, "##FIRSTX=";: Print#2
, using "#####. ###"; FIRSTX
        If Left$(Instrument$, 6)="PERKIN" then Print#2, "##FIRSTX=";: Print#2
, using "#####. ###"; LASTX
        Print#2, "##FIRSTY=";: Print#2 , using "#####. ########"; FIRSTY
        If Left$(Instrument$, 7)="Digilab" then Print#2, "##LASTX=";: Print#2
, using "#####. ###"; LASTX
        If Left$(Instrument$, 6)="PERKIN" then Print#2, "##LASTX=";: Print#2
, using "#####. ###"; FIRSTX
        Print#2, "##XFACTOR= 1"
        If Left$(Instrument$, 6)="PERKIN" then Print#2, "##YFACTOR=";: Print#2
, using "###. ########"; YFACTOR
        If Left$(Instrument$, 7)="Digilab" then Print#2, "##YFACTOR= 0. 0001"
        If Left$(Instrument$, 6)="PERKIN" then Print#2, "##DELTAX= ";: Print#2
, using "##. #####"; XDataVal(1)-XDataVal(2)
        If Left$(Instrument$, 7)="Digilab" then Print#2, "##DELTAX= ";: Print#2
, using "##. #####"; XDataVal(2)-XDataVal(1)
        Print#2, "##NPOINTS="; NOPOINTS%
        Print#2, "##XYDATA= "; XYDATA$
        Temp$=""
        Temp1$=""
        Temp2$=""
        Temp3$=""
        Locate 17, 5: Print
"Please wait I am writing your data to file "; NewFileName$
        Locate 18, 5: Print "Number completed =       =        %"
        If Left$(Instrument$, 6)="PERKIN" then
                I1%=NoPoints%
                56 Temp$=Str$(INT(XDATAVAL(I1%)))
                Temp$=Mid$(Temp$, 2, Len(Temp$))
                for i2%= 1 to 10
                         Temp1$=Str$(int(YDATAVAL(I1%)))
```

```
                                Temp1$=" "+Temp1$
                                Temp2$=Temp2$+Temp1$
                                if i1%=1 then i2%=10
                                Locate 18,23: print NoPoints%-I1%+1
                                Locate 18,32: print
using "###.#";((NoPoints%-I1%+1)/NoPoints%)*100
                                I1% = I1% - 1
                    next I2%
                    Temp3$=Temp$+Temp2$
                    Print#2, Temp3$
                    Temp$=""
                    Temp1$=""
                    Temp2$=""
                    Temp3$=""
                    if i1%=0 then 57 else  56
                    57 Print#2,"##END="
                    Close#2
          ElseIf Left$(Instrument$,7)="Digilab" then
                    I1%=1
                    61 Temp$=Str$(XDATAVAL(I1%))
                    Temp$=Mid$(Temp$,2,Len(Temp$))
                    If Mid$(Temp$,4,1)="." then Temp$=Left$(Temp$,7)
                    If Mid$(Temp$,5,1)="." then Temp$=Left$(Temp$,8)
                    for i2%= 1 to 10
                                Temp1$=Str$(int((YDATAVAL(I1%)*YFactor)/0.0001))
                                Temp1$=" "+Temp1$
                                Temp2$=Temp2$+Temp1$
                                if i1%=NoPoints% then i2%=10
                                Locate 18,23: print I1%
                                Locate 18,32: print using "###.#";((I1%)/NoPoints%)*100
                                I1% = I1% + 1
                    next I2%
                    Temp3$=Temp$+Temp2$
                    Print#2, Temp3$
                    Temp$=""
                    Temp1$=""
                    Temp2$=""
                    Temp3$=""
                    if i1% NoPoints% then 60 else 61
                    60 Print#2,"##END="
                    Close#2
          End If
          Gosub Returnret:
          Return

DrawSpectra:
          If Left$(Instrument$,7) = "Digilab" then GOSUB DigilabDataLoad:
          If Left$(Instrument$,7) = "Mattson" then Gosub MattsonDataLoad:
          If Left$(Instrument$,6) = "PERKIN" then Gosub PerkinDataLoad:
          Gosub Returnret:
          Cls
          Screen 8
          color 1,0
          If Left$(Instrument$,7) = "Digilab" then GOSUB DrawDigilab:
          If Left$(Instrument$,7) = "Mattson" then Gosub DrawMattson:
          If Left$(Instrument$,6) = "PERKIN" then Gosub DrawPerkin:
          30 x$=INKEY$
          If x$  " " then Locate 25,1: goto 30
          Gosub Returnret:
          cls
          Return

DrawPerkin:
          XStep=XDataVal(1)-XDataVal(2)
```

```
        x1=SmallestPeNo-(SmallestPeNo*0.1)
        x2=LargestPeNo+(LargestPeNo*0.1)
        window (LastX,x1)-(FirstX,x2)
        Pset (LastX,YDataVal(1))
        X=LastX
        For I%= 2 to NoPoints%
                X=X+XStep
                Line -(X,YDataVal(I%))
        Next I%
        Return

DrawDigilab:
        XStep=XDataVal(2)-XDataVal(1)
        If SmallestDiNo  0 then x1=SmallestDiNo+(SmallestDiNo*0.05)
        If SmallestDiNo = 0 then x1=SmallestDiNo-(SmallestDiNo*0.05)
        x2=LargestDiNo+(LargestDiNo*0.05)
        window (FirstX,x1)-(LastX,x2)
        Pset (FirstX,YDataVal(1))
        X=FirstX
        For I%= 2 to NoPoints%
                X=X+XStep
                Line -(X,YDataVal(I%))
        Next I%
        Return

DrawMattson:
        XStep=XDataVal(2)-XDataVal(1)
        x1=SmallestMaNo-(SmallestMaNo*0.1)
        x2=LargestMaNo+(LargestMaNo*0.1)
        window (FirstX,x1)-(LastX,x2)
        Pset (FirstX+0.0001,YDataVal(1))
        X=FirstX
lprint "xstep";xstep,"x1";x1,"x2";x2,"Firstx";Firstx,"Lstx";Lastx,"x";x
        For I%= 2 to NoPoints%
                X=X+XStep
                Line -(X,YDataVal(I%))
        Next I%
        Return

Border:
        Screen 8
        color 3,0
        Locate 1,1:
print "  IMMMMMMMMMMMMMMMMMMMMMMMMMMMMMMMMMM
MMMMMMMMMMMMMMMMMMMMMMMMMMMMMMMMMMMMMMMMMM; "
        For I%= 2 to 22
                Locate I%,1:print " :":Locate I%,78:Print ":"
        Next I%
        Locate 23,1:print "  HMMMMMMMMMMMMMMMMMMMMMMMMMMMM
MMMMMMMMMMMMMMMMMMMMMMMMMMMMMMMMMMMMMMMMMMMM"
        Return

Title:
        Locate 7,34: Print "JCAMP FILE"
        Locate 9,25: Print "DISPLAY & CONVERSION PROGRAM"
        Locate 11,18: Print "Perkin-Elmer  Digilab  Mattson"
        Return

Returnret:
        color 20,0
        locate 25,34: print "SPACE BAR ";
        color 9,0
        locate 25,27: print " Press ";
        locate 25,44: print "to continue ";
```

```
8 x$=inkey$
if x$   " " then locate 25, 56: goto 8
cls
return
```

ERRORTRAP:
```
If ERL= 80 or ERL = 81 then Resume Next
If Err = 53 then
        Color 20, 0
        Locate 12, 17: PRINT "NO FILES FOUND ..... Please Press <RETURN>
"
        Color 2, 0
        er=53
        Resume Next
End If
print err, erl
stop
```

CHAPTER 12

User Written Software

I. STEER

With the ready availability of personal computers and of sophisticated high level languages there is a tremendous opportunity to write software for one's particular requirements. Before embarking on such a task, there are some important points to consider.

1   WIIY WRITE ONE'S OWN SOFTWARE?

In most cases, particularly for scientific work, commercial software packages are written to satisfy the largest market. The consequences are that the software product may not do exactly what the user wishes but stops about 95% of the way. This phenomenon is difficult to avoid as each individual customer has his or her own specific needs and to try to satisfy all needs would result in a massive unwieldy product that is difficult to use. By the nature of the extent of the product the system would also be expensive.

In order to try and optimize the applicability of a particular package, manufacturers tend to design the product to consist of a series of building blocks that can be operated in any order to provide the flexibility.

There is therefore an interest in writing a program to add the extra items needed to complete the task.

Many scientific instruments can output data via an RS-232-C interface directly to a computer or printer. In a major proportion, the instruments can be controlled directly by commands sent along this RS-232-C interface. The opportunity arises to write a small program to control the instrument and collect data without having to purchase a commercial package.

Once a program has been fully tested and debugged, that program should continue to perform as accurately as ever time and time again. The use of such a program makes a task much easier than trying to manually transcribe results from an instrument. Human error on average produces about 2% of errors, and if the task becomes tedious then this error rate will climb.

There is no doubt that it is worth writing one's own programs, but one must be aware of the pitfalls and dangers.

1.1   Dangers

The most important consideration is the performance of the software and in particular the safety aspects.

If numerical results are generated by the software for a quantitative measurement then it is important that the software is tested for correct results over the whole number range of the measurement and for over and under range conditions. Rounding errors that may happen during the calculation must be considered. In this instance the actual precision to which the calculation is performed must also be considered at the start of program design.

In this respect it is actually safer if the software crashes (*i.e.* fails to continue) rather than produce an incorrect result. It is worth noting the requirements of Good Laboratory Practice in your working environment.

The importance of testing software packages can be seen in the fact that many more hours are spent testing the software package than actually generating the code for a commercial package.

For a simple program designed to do a specific job for a limited number of users there is less requirement for the program to cope with such items as incorrect key presses or incorrect entries. It may therefore be perfectly acceptable that the program actually stops when such an error occurs.

Another common problem with user written software is that the project snowballs and no final program is obtained in a reasonable time. This is usually due to the programmer becoming too interested in side issues, and becomes totally distracted from the actual task. This is bad project management and the writer must exercise control to ensure that the project keeps to its original aims. Any deviation should be evaluated to consider its usefulness and the implications on the project time-scale considered. If a project is not managed in this way it will appear to a manager that the job is not viable and time has been wasted. The possibility of persuading the management that it is worth continuing the project or starting a new task becomes an almost impossible job. This means that a golden opportunity has been wasted.

1.2   What Does One Want the Programs to Do?

User written programs can be divided into approximately three categories;
(i)   Collect data directly from an instrument. (This might include instrument control but might only be for printer output.)
(ii)   Collect data from an instrument into a commercial program.
(iii)   Directly access data stored by a commercial package which controls the instrument and generates the data. (Then write additional program for data manipulation.)

Let us now look at the advantages and disadvantages of each method.

1.2.1   Collecting Data Directly from an Instrument. The ease of using this approach depends directly on the degree of sophistication built into the instrument. Should the instrument contain an RS-232-C interface and allow data transfer as ASCII characters then the task is really rather simple. Instruments to which this approach is applicable are the Philips PU8620 series of UV/Visible spectrometers.

If an instrument has no such output, but has only an analogue voltage for the signal and requires control lines to be toggled to provide control, then this is not a simple task. An example of such an instrument is the PU9700 Infrared Spectrometer. Philips have manufactured an Analogue to Digital Converter printed circuit board (ADC) for an IBM compatible computer which controls this instrument. The ADC digitizes the analogue output for the computer. Control lines are connected to the board to allow the computer to monitor the stepper motor drive of the instrument and so control the instrument and monitor the wavenumber position of the instrument.

Stepper motor pulses used to trigger data collection can be output from the instrument at up to 150 Hz. During this time the software must:

(a)   Read the ADC,
(b)   Convert the reading into true % Transmittance,
(c)   Store the value in the computer's memory,
(d)   Update the screen display so the spectrum is drawn on the screen as the scan progresses,
(e)   Check to see if the special function key on the computer has been pressed to abort the scan,
(f)   Be ready for the next pulse.

In order to have a program to operate fast enough the program must be written in ASSEMBLER. Even so the updating of the display may take so long that data points can be missed. To circumvent this problem Direct Memory Access (DMA) is used to load the ADC output automatically into a scratch pad area in the memory of the computer. The ASSEMBLER language program then maintains a supervisory role and acts on a data point when one has been loaded into the scratch pad area of memory.

The supervisory program is therefore responsible for stages (b) to (f) above. When the scan is stopped the supervisory ASSEMBLER program can determine exactly how many data points have been collected by the DMA so no information is lost.

Unfortunately this is not the end of the story since there may be periods in which the data output from the instrument is invalid, for example, dead periods in the scanning of the spectrometer when the instrument is changing gratings. The ASSEMBLER program must therefore ignore these readings and not update the screen nor load these invalid points into the data array.

It is clear from this summary of the required operations that to write such a program requires considerable knowledge of the instrument. It also requires considerable testing and a specially built ADC board. This particular ADC board was built specifically to provide the necessary ports to handle the instrument control lines. Such a task would not be considered suitable as a project for user written software.

1.2.2   Collecting Data from an Instrument into a Commercial Software Package. There are commercial packages available to connect directly to the analogue output of an instrument. However when one considers such problems as dead periods in IR scanning, this solution may not be ideal. However the application of such a package to kinetics measurements at a fixed wavelength may be ideal.

To use a commercial package such as LOTUS 123 or FRAMEWORK one needs to learn how to drive the package and to learn the programming language. If one is already familiar with the package or has routines already available then there may be considerable advantages in using this route. If not, the added cost of the package and unknown complexity may mean that a complete program written in BASIC may be more easily generated, tested and supported.

To summarize; the requirement is to match the software package to the application task and the instrument. To make such a judgement requires detailed knowledge of the workings of the instrument as there may be hidden snags and the task may multiply in difficulty.

Several examples of the use of these software packages exist.
(a) The Philips PU9200/9400 Atomic Absorption instruments output data through an RS-232-C interface. Results may be collected directly by using the communications facilities in FRAMEWORK.
(b) A User Programming software package is available for the PU8700 UV/Visible spectrophotometer. This package runs on the Personal Computer and enables a text file of commands to be generated. This command file can then control the instrument to collect and format the data and then invoke LOTUS automatically to carry out the required calculations. The program (or MACRO) to carry out the calculations must have been written using LOTUS.
(c) Several packages written by Philips for their instrumentation have the ability to generate a results file as a string of ASCII characters. This ASCII file may then be imported directly into LOTUS. What is then required is that the operator formats the data into the required spreadsheet cells ready for the calculation. The calculation routine must obviously be written in the LOTUS package.

The main advantages accrue from the operator already being familiar with LOTUS or similar packages. If the user is not familiar with the package then the extra cost of LOTUS, *etc.* plus the problem of setting up the program make this option less interesting as the job can be programmed more easily in BASIC. However some of the graphical capabilities of packages may sway the balance in the opposite direction.

1.2.3   Access Data stored on Disc from Commercial Instrument Control Packages. For many packages the data format for the information on disc is readily available from the instrument manufacturer. Many software packages for Philips instruments in IR, UV, and AA have a data format that may be readily picked up by BASIC, PASCAL, or C. The user need only write a program to provide the extra facility that is wanted. It may well be possible to re-examine the modified data with the commercial package.

It is strongly recommended that where the data are modified by such programs that the original raw data remain available (*cf.* Good Laboratory Practice).

The great advantage of this type of user programming is that the programming task is usually small. The main disadvantage is that the programs cannot be run as an entity as the data must be collected using the commercial package before then the user written program is run.

Having looked at the three possible categories for user written software, what option is the best?

The answer to this question depends on the capabilities offered by the instrumentation, the software packages currently available and the actual final task.

Another consideration is the knowledge and capabilities of the person who is to do the programming. This involves knowledge of programming languages, any commercial software such as LOTUS, and of interfacing instruments to computers.

Depending on the final aim, it is possible to plan various routes using different tools to obtain the goal. When each of these routes has been chosen the task is then to identify for each route the possible problems and therefore the risk. Some short investigations may be needed to clarify some areas of uncertainty. It is only then that a choice of option can be made.

It should not be forgotten that, whether in industry or academic life, the programs that are written by the user will be used as tools to get or modify data. It is therefore important that the time taken to write and test the program is as short as possible and that the software writer should not be distracted from the main task by interesting sidelines.

2 CHOICE OF PROGRAMMING LANGUAGES

Dependent on the task in hand, the choice of programming language may not be important. As time is of the essence, it may be best to use the language to which the software writer is accustomed.

Questions that should be asked are;
(i) Will the language provide all the facilities required for the program?
(ii) Is there any program already written that provides routines that are wanted in this program? If so then the programming language used for the earlier program is the one to use.
(iii) Is the current program likely to be used as a basis for future programs? If so then a language or version of the language should be chosen that gives the flexibility for later expansion.
The most commonly used languages are;
  BASIC
  PASCAL
  C
  ASSEMBLER

Mixed Language Programming

For each of those languages there are specialized forms which may have extra facilities available over the more standard version of the language. These forms are very powerful and produce programs that can be fast in operation.

However there are possible difficulties in their use in that it may not be possible to use them with, for example, a library of display driver routines or the programs generated may not be easily updated in the future so may not be quite as flexible as the general versions of the language produced by Microsoft.

However as these specialized forms become still more powerful, that form may be absolutely ideal for a single project. If software is to be written that can be transferred between systems and easily updated with the minimum of effort in rewriting code then it is best to use 'standard' languages from Microsoft.

2.1   The BASIC Language

BASIC is unfortunately considered to be a 'toy' language by many programmers who have been trained primarily as software writers. This statement was probably true of early versions of BASIC before the arrival of the Personal Computer.

BASIC is now an extremely powerful language with excellent graphics support and buffered RS-232-C communications facilities built into the language.

Normal BASIC is interpretive. That means that as the program is run each program statement is translated individually into code that the computer can understand and then action. The result is that the program runs relatively slowly. It also means that the BASIC language must be loaded into memory before the program can be run.

To circumvent this problem, BASIC compilers are available which take the original BASIC program and convert it into computer readable form. This compiled program can then run directly without the need to load BASIC. A speed increase of about a factor of 5 is obtained by compilation.

Besides graphics and RS-232-C capability, it is extremely useful to be able to use a Maths Coprocessor integrated circuit which provides high speed, high accuracy floating point mathematics. The IBM BASIC COMPILER Version 2.0 does not contain support for a Maths Coprocessor, but many other BASIC compilers, such as Microsoft Quick Basic and Borland Turbo Basic, do provide support.

Many of the newer versions of BASIC, such as QuickBASIC actually compile each line of program as it is entered and so the programs always run at compiled speed.

Even with compiled BASIC, the actual speed of operation of the program is not the fastest possible. For programs that require very rapid response such as for the PU9700 Infrared system mentioned earlier then the time-critical routines must be written in ASSEMBLER and called from the BASIC programs.

The major advantage of BASIC is that it is an easy language both to learn and to get a program running. If a program is required in the fastest possible time then BASIC is probably the answer. There are some limitations in BASIC such as easily accessing a data space of greater than 64K but this is rarely a problem for normal user written software.

2.2   The PASCAL Language

PASCAL was originally developed by Professor Wirth for teaching purposes. It was intended to show the benefits of a logical structured approach to programming. In its initial form the ability to handle strings of characters and its disc accessing capability were extremely limited due to the origins of the language. Even later versions had very limited graphic and communication capability. All PASCAL versions are compilers with no interpretive versions. The code generated by these compilers does not appear to be particularly efficient. In fact programs written in compiled BASIC can run faster presumably due to more efficient compilers.

Borland Turbo PASCAL has good graphics capability and does support RS-232-C communications. Version 5 of this compiler is very comprehensive providing an overlay manager and Expanded Memory Support.

PASCAL used to be THE language for programming particularly for scientific work. Today more and more programs are being written in C rather than in PASCAL.

2.3   The C Language

This is an extremely powerful language developed by Denis Ritchie in 1972. The design of the language leads to a structured approach to programming. C is extremely flexible and produces fast and efficient code. It is understandable that C is probably the most used language in the scientific community (PC environment) and many compilers are written in C. Writing programs for the WINDOWS environment or for PRESENTATION MANAGER virtually demands the use of C. Most of the modern C compilers support the presence of a Maths Coprocessor or emulate the presence of a Coprocessor. Simple graphics capabilities are available with some compilers, but buffered RS-232-C support is not provided. This situation is dramatically improved by the wealth of commercial libraries of routines available to provide these facilities.

In terms of usage C source code is more difficult to debug than that for other languages. This situation is improved by products like CODEVIEW from Microsoft which enable the source to be debugged as it is run.

One of the major advantages of C is the ease with which direct calls can be made to the Disc Operating System (DOS) or the Basic Input Output System (BIOS) of the computer for providing non-standard features. The method of calling routines written in ASSEMBLER Language from C is also particularly easy.

Because it is possible to write a single line of C code to do a number of tasks it is easy for the programmer to write code that another programmer would find difficult to understand.

The compiler probably most used is Microsoft C or Quick C. The newer versions are very comprehensive. The Zorland C compiler from Zortech is also very powerful with some different facilities.

2.4   The ASSEMBLER Language

This language provides access to the computer at the lowest level. As such the code produced is likely to be the fastest and most space efficient of all. Since ASSEMBLER is a low level language, the facilities offered are less than other languages and it is more difficult to program in this language than in a high level language.

The advantages of this language are the speed and space efficiency of the code. Again it is possible to access low level facilities in the computer BIOS and memory very easily.

The disadvantages of the language are the difficulties of programming and debugging in ASSEMBLER.

In general, ASSEMBLER is useful in cases that require either the speed of operation or the ability to access low level facilities not available in high level languages.

2.5   Mixed Language Programming

This is a facility available with the Microsoft suite of languages, ASSEMBLER, BASIC, C, FORTRAN, and PASCAL, but this capability is only available with the newer versions of the languages.

A slight complication arises with the use of BASIC in that the actual computer working environment is initialized in a special way when BASIC is invoked. This means that the program must be started in the BASIC environment whereupon calls can then be made to routines written in other languages such as C. It is not possible to have the main program written in say C and then call routines written in BASIC.

The major advantage of mixed language programming is the ability to add features to an existing program that may be more easily programmed or only available in another language. Should one already have a program written say in BASIC and wish to convert to C then one is faced by a major task of rewriting AND debugging. With mixed language programming it is possible to change or add one routine at a time. It is therefore easier to test and debug the rewritten routine and the work-load to convert to C can be spread over a longer time period.

Most high level languages allow calls to be made to ASSEMBLER routines. The Borland Turbo languages are usually excellent in this capability allowing in-line ASSEMBLER statements.

2.6   Summary

What is the best language?

(i)   Use the language to which the programmer/user is accustomed if it will do the job.
(ii) The easiest and quickest route is to use BASIC.
(iii) Probably the outstanding performer in terms of facilities and speed of code operation is C.
(iv) For highest speed of operation use ASSEMBLER.

3   THE PHILOSOPHY IN WRITING PROGRAMS

3.1   Keep it Simple!!

This is THE key philosophy. Programs that are kept simple are easier to understand, modify, and maintain. Because the simpler program uses less code it is likely to run faster. Esoteric routines that may do wonders for the software writer's ego, but nothing in terms of functionality of the program must be avoided.

Any software package will need to be maintained or referred to for further extension. The source code must be examined and understood again some weeks or months after the original coding. A simple program with an adequate level of annotation is necessary to ensure that it can be understood by another software writer at some future time. A program may seem obvious to the writer at the time, but in a few month's time this is rarely the situation.

If the writer discover idiosyncrasies or undocumented features of a compiler then these should be noted carefully for future reference otherwise time will be spent rediscovering these facts a second or third time.

3.2   Plan the Project

Any project requires a clearly defined aim. The major tasks required to achieve this aim can be listed as a sub-set which should be followed by a further sub-set of the individual components of each of the tasks.

The above actions define a skeleton specification of the project. This specification should be written down in a formal notebook for reference during the project and for maintenance purposes.

The next stage is to examine how the project can be achieved. This means examining each component task to determine how it may be accomplished. Any unknowns such as 'Will data collection be fast enough' must be identified and highlighted as risk areas should exist.

At the end of this planning stage a breakdown of the component parts of the project with the high risk areas should exist.

The importance of identifying the high risk areas is crucial as this will ultimately determine the viability of the project. From this plan it will now be possible to estimate time scales for writing and testing the programs. This breakdown can be used so that the programmer can gauge how well the project is progressing and can be shown to a supervisor so that he/she can see the progress. In this context the viability of the project must be assessed by drawing up a work plan to tackle the high risk areas as a priority. This effort will enable more accurate time estimates to be made and may indicate that a total redesign is needed.

It cannot be stressed how important it is to have a simple layout of the task in hand. Perfectly viable projects have been halted because the writer cannot show how progress has been made and the impression given to managers or supervisors was that the time and efforts had been wasted. The likelihood of agreement being given for any user programming again will be small.

If any component part of the plan appears too large then it should be broken down into smaller parts. Careful control on time spent on each component part is essential. If it looks that the project is getting bogged down then the situation must be reassessed by asking the following questions.

Is there information that can help?

Is there another way of doing the job?

It may well be that a particular project has to be abandoned. If, however, can be seen that time was spent correctly and the project was well managed then this was the correct decision. Although the project was abandoned it has shown both the potential for some future work and the project management capability of the software writer.

The whole viability of the project rests on exacting a successful conclusion for areas of high risk. It is therefore crucial that these problems are tackled first so that time is not wasted.

An Example Program

In order to see this philosophy in action, let us look at an example. The requirement is to collect data from a PU 9500 Series Infrared spectrometer and to display the data after data collection. It would be preferable to display the data as the scan progresses, but this may take too much time and may be too complicated to attempt at first. It will always be possible to modify the final program to produce a live display.

It is probably worth emphasizing that it is a much easier task to write a program to modify data that has been stored by a previous commercial package. Less experienced readers may prefer to jump to that section before attempting to understand this more complicated example of programming.

What is the Aim?

To generate a program to control the PU9500 Infrared spectrometer. To collect data from a scan and display the scan at the end of data collection.

What are the Major Tasks?

(i)   To connect the PU9500 to the computer.
(ii) To send a scan command to the spectrometer.
(iii) To collect data into the computer memory.
(iv) To display the collected data.

The first task is a hardware problem but assistance is provided by means of information in the instrument and computer user manual.

The PU9500 spectrometer has an RS-232-C interface and can be controlled by, and output data as, ASCII characters. Connecting the RS-232-C interface of the computer is usually relatively easy providing simple rules are followed. To test that the RS-232-C interfaces have been correctly connected, a simple program to send a scan command to the computer can be used.

What are the High-Risk Areas?
(a) Controlling the spectrometer because this is a totally unknown task.
(b) Can data be collected from the instrument and into memory fast enough before the next line of data is ready?
(c) Does the language have suitable graphics capability?
(d) Does the language support buffered RS-232-C communications or does a routine for communications exist that could be used?

The language chosen in this text for this example is IBM BASICA or Microsoft GWBASIC. This language has adequate graphics and RS-232-C support. The two remaining high risk areas are therefore (a) and (b). Of these two problems, (a) is obviously the most critical because if this fails the project is dead.

3.2.1. Controlling the Spectrometer. This is the critical element. There is the problem of making the connection from the spectrometer to the computer. For a standard IBM Asynchronous Communications card (RS-232-C), the null modem cable supplied with the spectrophotometer can be plugged in directly between the computer and the spectrometer. All the RS-232-C settings such as baud rate and parity should be checked on the instrument.

If the computer has more than one RS-232-C card then these cards must be checked to ensure that they are set to COM1 and COM2.

The setting of COM1 and COM2 defines the physical address that the computer uses to drive the card. If both cards are set to the same address then NEITHER RS-232-C will function. This problem also applies to printer cards with LPT1 and LPT2. Many non-working systems have been found to fail because of the above problems particularly if multi-function cards are used.

Now to the programming. What is involved to control the spectrophotometer? These stages again can be listed by reference to the instrument and BASIC language manuals.
(1) Open communications to the spectrometer.
(2) Send an ESCAPE character to take control of the spectrometer.
(3) Read in hyphen prompt sent in response by instrument.
(4) Send SCAN command to spectrometer.
(5) Read in scan header consisting of 4 lines of text and display on screen. (This will confirm communications are correct.)
(6) Send ESCAPE character to abort the scan then the KEYBOARD command to relinquish control to spectrometer keyboard.
(7) Close communication channel.

These stages can now be converted into a program.
```
10    OPEN "COMl:9600,E,7,1" AS #1
20    REM OPEN COMMS 9600 BAUD EVEN PARITY 7 DATA BITS 1 STOP BIT
```

```
30    PRINT #1,CHR$(27)
40    REM SEND ESCAPE CHARACTER TO SPECTROMETER
50    LINE INPUT #1,A$
60    REM READ IN HYPHEN PROMPT. A$ SHOULD CONTAIN THE HYPHEN PROMPT
70    PRINT #1,"SCAN"
80    REM SEND SIMPLE SCAN COMMAND TO INSTRUMENT
90    FOR X%=1 TO 4:LINE INPUT #1,A$:PRINT A$:NEXT
100   REM READ IN 4 LINES OF DATA HEADER AND PRINT ON COMPUTER SCREEN
110   PRINT #1,CHR$(27):PRINT #1,"KE"
120   REM SEND ESCAPE CHARACTER TO ABORT THE SCAN THEN KEYBOARD
      COMMAND TO RETURN CONTROL TO INSTRUMENT KEYBOARD.
130   CLOSE #1
140   REM CLOSE COMMUNICATION CHANNEL.
150   END
```

This small program will start a scan on the spectrometer, display the data header on the computer screen, then abort the scan.

A possible difficulty that can occur using RS-232-C interfaces is as follows. The PU9500 buffers data in its RS-232-C interface as does the computer. If the program fails to work correctly and just stops then data may be left in either of these buffers. The only absolutely certain method of clearing these buffers is to switch off both the computer and instrument for at least 15 seconds before restarting. It should only be necessary to resort to this process if the program fails to work.

Having succeeded in controlling the spectrometer, the only remaining problem is ensure that data can be collected at a fast enough rate from the spectrometer. The program therefore needs to be extended to collect data from the spectrometer. To do this the format of the data sent from the instrument must be known (Figure 1).

```
Line  no. 1234567891012345678920123456789301234567894012345678950123456

                                                                      CL.
      1    4000.0    −100.00      95.35     95.01    86 .75    81. 29RF
                                                                      CL.
      2    3980.0      80.26      74.29    −70.79    79 .00    85. 23RF
                                                                      CL.
   Last     200.0      11.00                                         1RF
                                                                      CL.
   Last     200.0      23.46      34.75                              2RF
                                                                      CL.
   Last     200.0      61.27      61.31     61.30                    3RF
                                                                      CL.
   Last     200.0      84.16      84.20     84.11    84 .11          4RF
                                                                      CL.
   Last     200.0      52.17      52.17     52.14    52 .13    52. 20 5RF
   End     −CL
            RF
```

Figure 1 *Data output format for* PU 9500

Data is transmitted from the spectrometer as lines of ASCII characters. Each line contains a wavenumber position followed by upto 5 intensity values (%T). The wavenumber position of the first intensity value is that at the listed wavenumber. The wavenumber positions of the remaining data points depends on the data interval. The character position in the line of each piece of data is constant so that it is easy to pick out the necessary information.

The last line of data may not have all five data points listed and it is unique in that the line is always longer than 53 characters and the number of data points in the line is given in character position 53.

End of data transmission is indicated by a hyphen prompt. This consists of the hyphen, a space character followed by the Carriage Return (CR) and Line Feed (LF) characters.

There is now an important point concerning the action of the LINE INPUT #1 statement in BASIC. This statement reads a line of data upto and including the Carriage Return character. Every line is terminated by a Carriage Return character followed by a Line Feed character. After the first line is read, the Line Feed character will remain in the RS-232-C buffer and will be read as the first character of the next line. The net effect is that all lines of data after the first line are shifted one character position to the right. This fact is important when attempting to extract data values from a line.

The method that LINE INPUT uses may not have been immediately obvious to the software writer. It is clear that testing of the data collection routine is important to ensure that there have been no unforseen mistakes.

3.2.2. Collect Data from PU9500 into the Computer Memory. What are the stages involved.
(1) Read each line of data.
(2) Is it the last line of data? If it is the last line of data then extract the required number of data points and finish.
(3) For each data point
  Extract value of data point.
  Scale data point to an integer. For %T data provided with 2 places of decimals the ideal scaling factor will be times 100. This will ensure that no precision is lost yet the scaling can cope with the largest and smallest %T value of +300%T and −300%T.
  Load an integer array with the extracted value.
(4) When all the data points have been extracted then for the last line of data it is only necessary to read the hyphen prompt which signals the end of data collection.
(5) Relinquish control of the spectrometer.

The simple program that has been outlined above has no method of checking for a successful completion of the task. The simplest method to ensure that all is well is to display the data which is actually the final stage of the program.

The program must be modified to carry out these tasks. Since an integer array is to be used to hold the data an integer array must be dimensioned of sufficient size to hold the data.

For a data interval of 1 cm^{-1}, data is output at 2 cm^{-1} above 2000 cm^{-1} and 1 cm^{-1} below 2000 cm^{-1}. For an instrument that spans 4000–200 cm^{-1} the number of data points is ((4000−2000)/2)+(2000−200)+1 or 2801 data points.

```
10      DIM D%(2801):SCAL%=100:COUNT%=1
20      REM DIMENSION INTEGER ARRAY OF 2801 VALUES. DEFINE SCALING
        FACTOR (SCAL%) =100. INITIALIZE DATA POINT COUNT TO 1.
30      OPEN "COMI:9600,E,7,1" AS #1
40      PRINT #1,CHR$(27)
50      LINE INPUT #1,A$
60      PRINT #1, "SCAN TIME 7 DATA 1"
70      REM SET FULL RANGE SCAN OF 7 MINUTES DATA INTERVAL 1 CM−1
80      FOR X%=1 TO 4:LINE INPUT #1,A$:NEXT
90      LINE INPUT #1,A$:IF MID$(A$,2,1)="−" THEN GOTO 230
100     REM READ IN A LINE OF DATA. IF THIS IS THE END OF DATA TRANSMISSION
        THEN GOTO LINE 230
110     MI!=VAL(MID$(A$,10,7))★SCAL%:D%(COUNT%)=MI!:COUNT%=COUNT%+
        1
120     REM EXTRACT THE FIRST DATA POINT FROM ITS POSITION IN THE LINE. MI!
        MUST BE A FLOATING POINT NUMBER BECAUSE THE VALUE OF THE DATA
        POINT IS FLOATING POINT. INCREMENT DATA POINT NUMBER.
130     IF LEN(A$)>54 AND MID$(A$,54,1)="1" THEN GOTO 90
140     REM IF LENGTH OF A$ IS GREATER THAN 54 THEN THIS IS THE LAST LINE OF
        DATA. THE NUMBER OF DATA POINTS IS GIVEN IN CHARACTER POSITION 54.
        IF THIS IS "1" THEN THIS IS THE LAST DATA POINT SO GOTO LINE 90 TO READ
        END OF TRANSMISSION LINE.
150     MI!=VAL(MID$(A$,19,7))★SCAL%:D%(COUNT%)=MI!:COUNT%=COUNT%+
        1
160     IF LEN(A$)>54 AND MID$(A$,54,1)="2" THEN GOTO 90
170     MI!=VAL(MID$(A$,28,7))★SCAL%:D%(COUNT%)=MI!:COUNT%=COUNT%+
        1
180     IF LEN(A$)>54 AND MID$(A$,54,1)="3" THE GOTO 90
190     MI!=VAL(MID$(A$,37,7))★SCAL%:D%(COUNT%)=MI!:COUNT%=COUNT%+
        1
200     IF LEN(A$)>54 AND MID$(A$,54,1)="4" THEN GOTO 90
210     MI!=VAL(MID$(A$,46,7))★SCAL%:D%(COUNT%)=MI!:COUNT%=COUNT%+
        1:GOTO 90
220     REM THERE IS NO NEED TO CHECK THIS LINE AS THE LAST LINE AS LINE 90
        WILL AUTOMATICALLY DETECT THE END OF TRANSMISSION LINE.
230     PRINT #1,"KE"
240     REM RETURN CONTROL TO 9500 KEYBOARD
250     CLOSE #1
260     FOR U%=1 TO COUNT%−1:PRINT D%(U%);:NEXT
270     REM WE HAVE COUNT%−1 DATA POINTS. LINE 260 PRINTS THE VALUES
        ACROSS THE SCREEN.
280     END
```

If a COMMUNICATION BUFFER OVERFLOW message is printed on the screen then we know that data is being generated too quickly. One cure may be to increase the size of the communications buffer in the computer. This is done when the BASIC interpreter is started. Should this fail then faster data collection methods are needed.

3.2.3. Graphics. It only remains to add a graphical display to the program. The simple IBM graphics monochrome screen in mode 2 has 640 elements (pixels) across the width of the screen and 200 vertically. The topmost left-hand screen element is 0,0 and the bottommost right-hand element is 639,199. In order to define a box which leaves room for annotation around the box, the left-hand margin can be set to 47 elements and the right-hand margin to 591. The top margin can be 15 and the bottom 159.

What is needed to do to draw the screen?
(1) Define coordinates for screen box.
(2) Set screen into graphics mode.
(3) Clear the screen.
(4) Draw box to contain spectrum.
(5) Draw spectrum starting at left-hand edge ensuring that the spectrum is correctly scaled.

The above requirements do not stop out of range data from being drawn on the screen. Such over range points could cause the numeric range to exceed an integer and so cause the program to crash. For this simple program no attempt will be made to cope with this situation and it could be rectified at a later date.

Note that due to the fact that the actual data interval between data points above 2000 cm⁻¹ is twice that below 2000 cm⁻¹ it has been necessary to convert these positions to values that provide a constant data interval. To convert from true wavenumbers to 'equivalent' wavenumbers requires the statements;

if wavenumber is less than or equal to 2000 then equivalent wavenumber equals the actual wavenumber
else
equivalent wavenumber=((wavenumber−2000)/2)+2000).

In order to draw the actual graph the stages involved are;
(1) Set the initial X and Y points to the origin of box. The 'line draw' function actually draws a line between two points. The first point on the graph should lie on the left-hand axis at the start wavenumber. In order to use a 'FOR NEXT' loop to draw the point at the start wavenumber, a dummy point is needed so that a line can be drawn from the dummy point to the start wavenumber (first real point) on the left hand axis. As this line should be invisible, the program should not actually draw it.
(2) For each point;
(3) Calculate the vertical pixel position on the vertical axis. The offset from the bottom of the display box is given by;

%Transmittance times the vertical span in pixels divided by the %T display range.

To determine the actual vertical position on the screen to draw to, this off-set must be subtracted from the vertical position of the bottom of the display.

(4) Calculate the horizontal point to draw to. The offset from the left-hand axis is given by;

The number of data points minus one (the first data point should lie in the axis) times the width of the screen in pixels times the data interval divided by the wavenumber span of the display.

(5) Draw a line between the current and previous points. Set previous X and Y coordinates equal to current coordinates.

(6) Next point.

Note that the data interval, start and stop wavenumber can be arranged to be entered by the user in further modifications to the program.

The program becomes;

```
10     DIM D%(2801):SCAL%=100:COUNT%=1:DISPLX%=47:
       DISPRX%=591:DISPTY%=15:DISPBY%=159:HEIGHTSPAN%=
       DISPBY%−DISPTY%:WIDTHSPAN%=DISPRX%−DISPLX%
20     REM DEFINE ARRAY AND COORDINATE POSITIONS OF DISPLAY BOX.
       (DISPLX%=DISPLAY BOX LEFT HAND X COORDINATE,
       DISPRX%=DISPLAY BOX X RIGHT HAND COORDINATE,
       DISPTY%=DISPLAY BOX TOP Y COORDINATE,
       DISPBY%=DISPLAY BOX BOTTOM Y COORDINATE,
       HEIGHTSPAN%=SPAN IN PIXELS OF Y AXIS OF BOX,
       WIDTHSPAN%=SPAN IN PIXELS OF X AXIS OF BOX.)
22     REM *******************************************************
24     REM SET UP SCREEN MODE AND DRAW DISPLAY BOX              *
26     REM *******************************************************
30     SCREEN 2:CLS
40     REM SET SCREEN 2(640X200 MONOCHROME GRAPHICS) AND CLEAR THE
       SCREEN
50     LINE(DISPLX%,DISPBY%)−(DISPRX%,DISPTY%),,B
60     REM DRAW DISPLAY BOX OUTLINE
62     REM *******************************************************
64     REM CONTROL SPECTROPHOTOMETER                            *
66     REM *******************************************************
70     OPEN "COMI:9600,E,7,1" AS #1
80     PRINT #1,CHR$(27)
90     LINE INPUT #1,A$
100    PRINT #1,"SCAN HIGH 4000 LOW 600 TIME 7 DATA 1"
110    REM SEND SCAN COMMAND TO SCAN FROM 4000 TO 600 CM−1 WITH A
       DATA INTERVAL OF 1.
120    FOR X%=1 TO 4:LINE INPUT #1,A$:NEXT
122    REM *******************************************************
```

```
124    REM COLLECT DATA                                                      *
126    REM ***********************************************************
130    LINE INPUT #1,A$:IF MID$(A$,2,1)="—" THEN GOTO 230
140    MI!=VAL(MID$(A$,10,7))★SCAL%:D%(COUNT%)=MI!:COUNT%=COUNT%+
       1
150    IF LEN(A$)>54 AND MID$(A$,54,1)="1" THEN GOTO 130
160    MI!=VAL(MID$(A$,19,7))★SCAL%:D%(COUNT%)=MI!:COUNT%=COUNT%+
       1
170    IF LEN(A$)>54 AND MID$(A$,54,1)="2" THEN GOTO 130
180    MI!=VAL(MID$(A$,28,7))★SCAL%:D%(COUNT%)=MI!:COUNT%=COUNT%+
       1
190    IF LEN(A$)>54 AND MID$(A$,54,1)="3" THEN GOTO 130
200    MI! = VAL (MID$(A$,37,7))★SCAL%:D%(COUNT%)=MI!:COUNT%=
       COUNT%+1
210    IF LEN (A$)>54 AND MID$(A$,54,1)="4" THEN GOTO 130
220    MI!=VAL(MID$(A$,46,7))★SCAL%:D%(COUNT%)=MI!:COUNT%=COUNT%+
       1'GOTO 130
222    REM ***********************************************************
224    REM RELINQUISH CONTROL OF SPECTROMETER                         *
226    REM ***********************************************************
230    PRINT #1, "KE"
240    CLOSE #1
242    REM ***********************************************************
244    REM THIS IS THE GRAPH DRAWING SECTION                          *
246    REM ***********************************************************
250    PREVX%-47:PREVY%=159:DATAINTERVAL%=1
260    REM LOCATE DUMMY POINT ON GRAPH TO ORIGIN.
       IN ORDER TO DRAW GRAPH RELATIVE TO WAVENUMBER POSITION NEED TO
       KNOW THE DATA INTERVAL BETWEEN DATA POINTS TO CALCULATE ACTUAL
       POSITION FROM THE DATA POINT NUMBER. THE SCAN COMMAND HAS
       REQUESTED A DATA INTERVAL OF 1 SO ONE HAS BEEN INSERTED AS A FIXED
       VALUE.
270    START%=3000:STOP%=600
280    REM THE ACTUAL START WAVENUMBER IS 4000CM—1. SINCE AT 1CM—1
       DATA INTERVAL THE ACTUAL DATA INTERVAL IS 2CM—1
       THE EQUIVALENT START POSITION FOR A CONSTANT 1CM—1 INTERVAL IS
       ((4000-2000)/2)+2000 OR 3000. THE STOP WAVENUMBER IS 600 CM—1.
270    FOR U%=1 TO COUNT%-1
280    REM NOW TO DRAW EACH POINT
290    MI!=D%(U%)★HEIGHTSPAN%/(SCAL%★100)
300    REM CALCULATE Y PIXEL POSITION OF FIRST DATA POINT.
       THIS IS THE INTENSITY [D%(U%)/SCAL%] MULTIPLIED BY THE NUMBER OF
       PIXELS PER INTENSITY UNIT [HEIGHTSPAN%/100] SINCE THE GRAPH IS TO BE
       FROM 0 TO 100%T.
310    IF MI!>HEIGHTSPAN% THEN MI!=HEIGHTSPAN%
320    REM IF DISPLAY TRIES TO DRAW OFF THE TOP OF THE BOX THEN LIMIT IT TO
```

THE TOP OF THE BOX.
330   IF MI!<0 THEN MI!=0
340   REM LIMIT FOR BELOW BOTTOM OF BOX
350   CURRENTY%=DISPBY%−MI!
360   REM ACTUAL PIXEL POSITION IS OFFSET FROM BOTTOM AXIS OF DISPLAY BOX.
370   CURRENTX%=DISPLX%+ ((U%−1)★(WIDTHSPAN%★DATAINTERVAL%)/(START%−STOP%))
380   REM CALCULATE HORIZONTAL POSITION FOR NEW POINT. THIS IS THE LEFTHAND AXIS PIXEL POSITION PLUS THE FOLLOWING: (DATAPOINTNUMBER−1)★(WIDTHSPAN IN PIXELS) ★DATAINTERVAL/(WAVENUMBER SPAN OF SPECTRUM)
390   IF U%=1 THEN GOTO 410 ELSE LINE(PREVX%,PREVY%) −(CURRENTX%,CURRENTY%)
400   REM DRAW LINE FROM PREVIOUS TO CURRENT POINT PROVIDING IT IS NOT THE FIRST POINT.
410   PREVX%=CURRENTX%:PREVY%=CURRENTY%
420   REM SET PREVIOUS POINT TO CURRENT POINT.
430   NEXT
440   REM NEXT POINT
450   END

This program can be considerably improved. It is intended purely to show the principles. The actual REM statements used in this example may be too long to fit on one line and will need to be split. The number of actual remarks put in the program itself needs to be a balance to ensure that another person can rapidly understand the workings of the program without going into further detail.

In this version of Interpretive BASIC there is not the possibility of calling named sub-routines. This means that is more difficult to separate the sections or modules of a program other than by lines of remarks or '★'. When named sub-routines can be used then a good description of the sub-routine can be provided at the beginning of the routine obviating the need for line by line comments.

3.3   Accessing Data Stored by Data Collection Packages

Many packages written for control and data collection for instruments store data in a format which allows the user to pick up the stored data from disc and to write his/her own specific software to carry out extra routines.

For example with Philips instruments, the format for the BIRD Infrared Software consists of a block of integers. The first 256 integers consist of a header containing a description of the data. The first data points starts at integer number 257. This data format allows the data to be readily loaded by a program written in BASIC, PASCAL, or C.

For PASCAL or C the first 7 bytes of the file should be ignored as they are a special descriptor required for BASIC.

As an example, a program to load a spectral data file and find the %Transmittance at 1600 cm^{-1} is adequate.

Aim:

To load a file from an Infrared data disc and to find the %Transmittance value at 1600 cm^{-1}.

Major Stages:

(a) Load the spectral data file into an integer array.
(b) Extract relevant descriptive data from file header.
(c) Locate 1600 cm^{-1} data point, if present.
(d) Print %T value.

3.3.1 Loading Spectral Data from Disc. The spectral data files are stored with DOS files names of the form XXX.DAT. "XXX" refers to a number, *e.g.* 001 or 014 that are shown in the library index page of the software (Figure 2).

The integer array size required is for 2801 data points plus a header of 256 integers. The data is saved on disc using the BSAVE (Binary SAVE) facility.

The program to reload the data file 001.DAT into an array D% is therefore;

```
10   OPTION BASE 1
15   REM SET NUMBER BASE TO 1 RATHER THAN ZERO.
20   DIM D%(3057)
25   REM DIMENSION INTEGER ARRAY OF 3057 VALUES.
30   BLOAD "A:001.DAT", VARPTR(D%(1))
35   REM BINARY LOAD FILE "001.DAT" FROM THE DISC IN DRIVE A STARTING AT
     THE ADDRESS OF THE FIRST ELEMENT OF THE ARRAY D%.
```

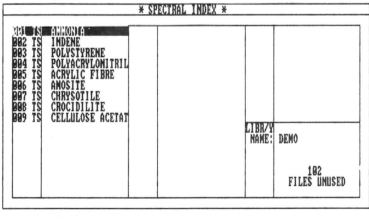

Figure 2 *Library index page*

3.3.2 *Extracting the Relevant Information from the Header.* The information needed is:
(1) Is the file valid? If D%(19)=78 then file is invalid.
(2) Start wavenumber. D%(2) gives the start wavenumber.
(3) Finish wavenumber. D%(6)
(4) Data interval. D%(106)/100
(5) Data format. (Abs or %T). D%(18)=65 then Abs.
    D%(18)=84 then %T.
(6) Intensity Scaling factor. A scaling factor of 100 is used for %T data and 10000 for all other data types.

The program becomes;
```
40    IF D%(19)=78 THEN PRINT "FILE INVALID" GOTO 120
45    REM IF FILE INVALID THEN PRINT MESSAGE AND QUIT.
50    IF D%(18)<>84 THEN PRINT "NOT %T DATA": GOTO 120
55    REM IF NOT %T DATA THEN PRINT MESSAGE AND QUIT.
60    IF D%(2)<1600 OR D%(6)>1600 THEN PRINT "NO DATA AT 1600":GOTO 120
55    REM CHECK SPECTRUM COVERS 1600CM−1
70    START%=D%(2):IF START%>2000 THEN START%=
      ((START%−2000)/2)+2000
75    REM CALCULATE START CM−1 IN EQUIVALENT CM−1
80    DATAINTERVAL%=D%(106/100)
85    REM LOAD DATAINTERVAL%. NOTE ONLY INTEGER DATA INTERVALS
      ALLOWED.
```

3.3.3 *Locating the 1600 cm^{-1} Data Point.*
```
90    RANGE%=START%−1600:A%=RANGE% MOD DATAINTERVAL%:
      IF A% >0 THEN PRINT "NO DATA POINT AT 1600":GOTO 120
95    REM CHECK IF DATA POINT AT 1600CM−1
100   NOPTS%= (RANGE%/DATAINTERVAL%)+1
105   REM CALCULATE DATA POINT POSITION OF 1600CM−1
```

3.3.4 *Printing the %T value.*
```
110   PRINT USING "###.## %T AT 1600CM−1",D%(NOPTS%)/100
115   REM PRINT %T VALUE
120   END
```

4   CONCLUSION

User written software can provide a rapid and easy route to obtain customized routines. Select the language and route to simplify the task. From the above examples it is obviously simpler to access stored data than to drive an instrument to collect the data. For some products this may not be possible.

Whatever route is chosen plan out the requirement and stages in the design. Look at the program and devise tests to check that it will work over all conditions.

Make sure that the program is adequately annotated.

The example programs provided in this section are given purely to illustrate points and are by no means complete programs, particularly for dealing with all eventualities and error trapping.

LOTUS 123 is a trademark of the Lotus Development Corporation.
FRAMEWORK is a copyright of Ashton Tate.
WINDOWS is a copyright of Microsoft.
PRESENTATION MANAGER is a product of Microsoft and IBM.
CODEVIEW is a product of Microsoft.

Subject Index